THE STRUCTURE AND EVOLUTION OF

NORMAL GALAXIES

North Atlantic Treaty Organisation, Advanced Study Institute held
at the Institute of Astronomy and Clare College, Cambridge
3 - 15 August, 1980

THE STRUCTURE AND EVOLUTION OF

N O R M A L   G A L A X I E S

edited by

S.M. FALL AND D. LYNDEN-BELL

Institute of Astronomy

University of Cambridge

Cambridge University Press

Cambridge

London - New York - New Rochelle
Melbourne - Sydney

Published by the Press Syndicate of the University of Cambridge

The Pitt Building, Trumpington Street, Cambridge CB2 1RP

32 East 57th Street, New York, NY 10022, USA

296 Beaconsfield Parade, Middle Park, Melbourne 3206, Australia

First published 1981

Printed in the United States of America

British Library cataloguing in publication data

The structure and evolution of normal galaxies.
1. Galaxies - Congresses
I. Fall, S. Michael   II. Lynden-Bell, D
523.1'12   QB857   80-42026

ISBN 0 521 23907 9

C O N T E N T S

CONTRIBUTORS

BALDWIN, J.E.   – Cavendish Laboratory, Madingley Rd., Cambridge.

BERTOLA, F.   – Osservatorio Astronomico, 35100 Padova, Italy.

BINNEY, J.J.   – Princeton University Observatory, Peyton Hall,
                      Princeton, N.J. 08540, U.S.A.

EKERS, R.D.   – Kapteyn Laboratorium, Groningen, Netherlands,
                      now at National Radio Astronomy Observatory,
                      PO Box 0, Socorro, New Mexico 87801, U.S.A.

FABIAN, A.C.   – Institute of Astronomy, Madingley Rd.,Cambridge.

FALL, S.M.   – Institute of Astronomy, Madingley Rd.,Cambridge.

FREEMAN, K.C.   – Mount Stromlo Observatory, Australian National
                      University, Private Bag, Woden P.O., A.C.T.
                      2606, Australia.

ILLINGWORTH, G.   – Kitt Peak National Observatory, 950 North Cherry
                      Ave., PO Box 26732, Tucson, Arizona 85726,U.S.A.

KORMENDY, J.   – Dominion Astrophysical Observatory, 5071 West
                      Saanich Rd., Victoria, B.C., V8X 3X3, Canada.

MADORE, B.F.   – David Dunlap Observatory, University of Toronto,
                      Box 360, Richmond Hill, Ontario, L4C 4Y6,Canada.

PAGEL, B.E.J.   – Royal Greenwich Observatory, Herstmonceux
                      Castle, Hailsham, Sussex, BN27 1RP.

SANCISI, R.   – Kapteyn Laboratorium, Postbus 800, 9700 AV
                      Groningen, Netherlands.

SCHWARZSCHILD, M.   – Princeton University Observatory, Peyton Hall,
                      Princeton, N.J. 08540, U.S.A.

TOOMRE, A.   – Dept. of Mathematics, M.I.T., 77, Massachusetts
                      Ave., Cambridge, Ma. 02139, U.S.A.

TREMAINE, S.   – Institute for Advanced Study, Olden Lane,
                      Princeton, N.J. 08540, U.S.A.

VAN DEN BERGH, S.   – Dominion Astrophysical Observatory, 5071 West
                      Saanich Rd., Victoria, B.C., V8X 3X3, Canada.

PREFACE

Detections of much greater gravitational attractions in the outermost parts of normal galaxies have led astronomers to believe that most large galaxies are surrounded by invisible haloes which may contain up to 90% of the total mass. These dominant masses of unknown dark material form a ghostly background to all discussions of galaxy formation and galactic structure, which are inevitably based on observations of the central 10% that shines. There, too, the new optical detectors have brought surprises along with improved data. The discovery that most elliptical galaxies rotate at half the rate that their ellipticities would indicate, and even less in many cases, has led to the suggestion that they are not oblate spheroidal but triaxial bodies slowly turning end over end.

It was to discuss the evidence for this new knowledge and its implications for theories of how galaxies form and evolve that a N.A.T.O. Advanced Study Institute was held in Cambridge, August 3 - 15 1980. This book contains an account of material presented in the more formal morning lectures given at the Institute. It contains a lucid account of current thought on both the dynamical structure of galaxies and their chemical evolution. By this we mean the way in which elements created in stars and supernovae are circulated back to enrich the interstellar and intergalactic gases and so to be re-incorporated into later generations of stars.

The introductory chapter by Fall gives an overall picture of the observable properties of galaxies. Theories of galaxy formation are designed to fit such facts into a coherent scheme In the following two chapters, Bertola and Illingworth discuss observations of elliptical galaxies, their distributions of brightness and colour and their spectra which are analysed to give their internal velocities and metal abundances. There are a number of galaxies in which the major axis of the isophotes twists through considerable angles as one proceeds to fainter brightness levels. Such twists can be quite readily explained if elliptical galaxies are triaxial bodies seen in projection. The bulges of spiral and SO galaxies have quite rapid rotations which make a striking contrast to the slow rotations of the elliptical galaxies. Dynamical models constructed to account for the latter are discussed by Schwarzschild and Binney. Stellar dynamical processes that can lead to one galaxy swallowing another are considered by Tremaine in a thought-provoking article that brings out the conundrums as well as discussing numerical experiments and physical explanations. Kormendy gives a broad review of recent work on barred spiral galaxies. It is believed that the bar pattern and the associated spiral structure rotate rigidly with a "pattern" angular velocity Ω_p. However, the "material" rotates at different velocities at different distances from the centre and the spiral is a density wave moving around it.

There is controversy as to whether the bars themselves are best regarded as regions in which the density wave is especially strong, or whether they are triaxial rotating bodies which are driving the spiral density wave in the rest of the galaxy. In most calculations based on the first picture, the stars that make up the bar move in distorted but near-circular orbits which crowd together in the region of the bar and so give rise to the density enhancement, although the individual stars cross over the inter-bar region. The alternative view is that the bar consists of stars on elongated orbits that never leave the bar itself. With new numerical experiments and much better observations of both gas and stars, a decision between these pictures is imminent.

Toomre's beautifully illustrated article provides here, for the first time, a unification of many ideas and much work on the generation of spiral structure in galaxies. It will be a surprise to many to see the great importance of very weak "leading" spiral waves which form the feed-back loop by which the strong trailing spirals are generated. Tidally generated spirals still find their place and may be responsible for many of the most spectacular examples seen in the sky. Baldwin considers the fine radio pictures of the two nearest and best observed spirals M31 and M33. It is less clear how these examples fit the idealisations of the theorist. Sancisi shows how such work extends to studies of the large scale features such as rings, warped disks and lopsided gas distributions, that are seen in the more distant spirals.

For some years past, the astronomical stage has been dominated by the fascinating phenomena displayed by active radio galaxies and quasars. These fall outside the scope of this book, but many normal galaxies show similar phenomena on a much muted scale. Ekers discusses the statistical properties of these non-thermal emissions of galaxies and leads us on to the surprising result that galaxies in interaction are more likely to show enhanced emission from their nuclei but not from their disks. Fabian's task is to open our eyes to the X-ray sky and to show how the rapid increase in sensitivity has changed our view of how galaxies interact with the hot gas that pervades the great clusters of galaxies. Any theory of galaxy formation must now contend with the iron observed in that gas, which tells us that it can not be primordial uncondensed matter, but is probably the product of supernova explosions and a long history both within and outside galaxies. Part of such a history is discussed by Van den Bergh who gives us a census of the nearest galaxies and discusses the behaviours of their interstellar gas clouds. Pagel gives an encyclopaedic picture of the chemical evolution of galaxies, and here we see recent observations from the Anglo-Australian telescope playing their role in the discovery that the barred spirals show less decrease of metal abundances from the centre to the bar end than do the similar parts of normal spirals. Crucial to all theories of the chemical evolution of galaxies is the mass birthrate function which gives the relative

numbers of stars of each mass born per unit time. Madore discusses
our rather fragmentary knowledge of this subject along with observations relevant to star formations.

Finally, Freeman discusses the distribution and motions of
the globular clusters which are perhaps the oldest constituents of
our Galaxy and finds further clues to the history of the way our
Galaxy formed.

To make this book as accessible as possible to those with
little formal training in astronomy, we have provided a glossary
with explanations of the more technical terms and abbreviations
used. This is to be found in front of the index.

We thank N.A.T.O. for sponsoring the Advanced Study Institute,
the contributors for the promptness of their manuscripts and
Professor Rees for his help on the organising committee.

D. Lynden-Bell

GENERAL PROPERTIES OF GALAXIES

S. Michael Fall

Institute of Astronomy, Cambridge

1 INTRODUCTION

The purpose of this chapter is to give a brief overview of
the general properties of galaxies, especially those that may be
relevant to an understanding of their origin and evolution.
Observations of the Milky Way and external galaxies are complementary
in this context because they give, on the one hand, an inside but
restricted view of a typical spiral galaxy and, on the other hand,
an outside but distant view of many galaxies. This body of knowledge is based on data taken in several different wavebands and
is now very extensive. In order to keep the exposition to a manageable length, it has been necessary to present some of the results
in highly abbreviated forms and to omit discussion of the methods
required to obtain them. For the same reason, no attempt has been
made to provide a complete set of references. More comprehensive
treatments can be found in the remainder of the book, for which
this chapter is intended as a general survey. In the following
discussion, the dependence of derived quantities on the extragalactic distance scale is indicated by the dimensionless Hubble
constant $h \equiv H/100$ km s^{-1} Mpc^{-1}; it is now thought to lie somewhere
in the range $0.5 \lesssim h \lesssim 1.0$. Unless otherwise stated, all optical
intensities pertain to the photographic B-band and all luminosities
are based on the B_T^O magnitude system, which includes corrections
for both inclination and galactic extinction.

2 CHARACTERISTIC MORPHOLOGIES

Although the taxonomy of galaxies is essentially subjective,
it does provide a useful framework for the quantitative description
of their properties. Most classification schemes are based on
Hubble's system, which recognizes four main classes: elliptical (E),
lenticular (SO), spiral (S) and irregular (I). E galaxies have a
spheroidal appearance with little or no evidence for internal structure and extinction. In addition to such a spheroidal or bulge
component, SO and S galaxies have a flat or disk component. The
distinction between these classes depends on the prominence of

1

spiral structure and HII regions in the disk, being less for S0 galaxies and more for S galaxies. Some I galaxies resemble disk galaxies but with less symmetry; others are designated 'peculiar' and often show signs of tidal interaction (Toomre & Toomre 1972). The relative frequencies of the main classes are roughly E : S0 : S : I $\simeq$ 13 : 22 : 61 : 4 for bright galaxies of the general field. In groups and clusters, the proportions of E and S0 galaxies are higher and, indeed, there is a rather tight correlation between the class frequencies and the local density of galaxies, with S and I galaxies almost absent in the densest regions (Dressler 1980).

The E class has been subdivided in a variety of ways but only two types require mention here. Members of the cD subclass are usually located near the centres of clusters and are distinguished from normal E galaxies by their extensive outer envelopes. The dwarf ellipticals, designated dE, are similar to normal ellipticals except for their smaller sizes and a much wider range in central surface brightness; like globular clusters, most of them are the companions of larger galaxies. The S0 and S classes are usually subdivided into families to indicate the presence or absence of bars. Because these features correlate only weakly, if at all, with the large-scale structure and content of disk galaxies, they are probably less fundamental than the classification schemes would suggest. One possibility is that both barred and ordinary spiral patterns are manifestations of the same dynamical process. Of greater significance is the relative prominence of the disk and bulge components, which provides one of the traditional criteria for the stage designations a, b, c, d from 'early' to 'late' spiral types. On average, the disk-to-bulge luminosity ratio increases from about one at Sa to about ten at Sd (Yoshizawa & Wakamatsu 1975) and this in turn correlates with colour and gas content. S0 galaxies appear to have an even wider range of disk-to-bulge ratio, with some values as low as 0.3.

3 INTENSITY PROFILES

The projected distributions of optical intensity in normal E galaxies and the bulges of S0 and S galaxies are similar and have been approximated by several fitting functions. The so-called $r^{\frac{1}{4}}$ law gives the intensity along an isophotal contour of area πr^2 as

$$I_B(r) = I_e \exp \{ -7.67 [(r/r_e)^{\frac{1}{4}} - 1] \}. \tag{1}$$

Here I_e is the intensity at the 'effective radius' r_e, which encloses half of the luminosity $L_B = 23 I_e r_e^2$. Near r_e, the logarithmic gradient of this profile is $- 1.9$, in close agreement with Hubble's law but significantly steeper than the isothermal profile. In general, L_B correlates positively with r_e but power-law fits to the data on E galaxies give a wide range of exponents, from 0.8 (Kormendy 1977) to 1.7 (Strom & Strom 1978). The projected shapes

Fig. 1. Bulge of the Milky Way. The horizontal extent of the field is $45° = 7$ kpc (D/9 kpc) and the axial ratio of the bulge is $b/a \simeq 0.8$. Taken by R.E. Royer with a 135 mm Schneider Symmar S lens at f 5.6 : 50 min exposure 098 emulsion, W1A filter.

Fig. 2. Bulge of NGC 4565. The horizontal extent of the field is $3'.0 = 16$ kpc (D/18 Mpc) and the axial ratio of the bulge is $b/a \simeq 0.6$. Taken by S.M. Fall and D.A. Hanes with the 3.9 m AAT at prime focus: 30 min exposure, IIaD emulsion, GG 495 filter.

of E galaxies have axial ratios b/a between 1.0 (E0) and 0.3 (E7) so the axial ratios of their three-dimensional shapes must lie between similar limits. In some cases, the images show signs of isophotal twists and radial changes in ellipticity, suggesting that their true figures are generally in the form of triaxial ellipsoids. But whether these are more nearly oblate, with one short axis, or more nearly prolate, with two short axes, is not yet known.

The distributions of intensity in the disk component of most S0 and S galaxies can be fitted by an exponential function:

$$I_D(r) = I_o \exp(-\alpha r), \tag{2}$$

where r denotes galactocentric distance. The luminosity of this profile is $L_D = 2\pi I_o \alpha^{-2}$, half of which is emitted inside a circle of radius $r_e = 1.67 \alpha^{-1}$. For reference, the Milky Way has a total luminosity $(L_D + L_B)$ of $L \simeq 1.6 \times 10^{10} L_\odot$, an effective radius of $r_e \simeq 5$ kpc, a disk-to-bulge ratio of $L_D/L_B \simeq 2$ and is classified as an Sb or Sc galaxy (de Vaucouleurs & Pence 1978). When the profiles of well-studied disks are extrapolated to their centres, the associated intensities are usually found to be near $I_o \simeq 145 \, L_\odot$ pc^{-2} (Freeman 1970); this implies a luminosity-radius relation of the form $L_D \propto r_e^2$ but it may, in part, reflect a preference for selecting galaxies in a limited range of surface brightness. Another useful parameter is the Holmberg radius, defined by the isophotal contour at $1.7 \, L_\odot pc^{-2}$; in most cases, one finds $r_H \simeq 4.5\alpha^{-1}$ and little evidence for spiral structure beyond this point. The intensity of the disk component falls off quite rapidly away from the equatorial planes of most edge-on galaxies and the corresponding scale heights are of order $0.1\alpha^{-1}$. In addition to this 'thin disk', some S0 galaxies appear to have a 'thick disk', which forms a more extended distribution of intensity at faint levels (Burstein 1979).

4 LUMINOSITY FUNCTIONS

When considering the properties of individual galaxies, it is sometimes useful to make reference to the general luminosity function. This is defined such that $\phi(L)dL$ is the average density of galaxies with total luminosities in the small interval $(L, L + dL)$ and it can be fitted by the function

$$\phi(L) = (\phi*/L)\exp(-L/L*), \tag{3a}$$

$$\phi* \simeq 1.9 \times 10^{-2} h^3 \, Mpc^{-3}, \tag{3b}$$

$$L* \simeq 1.0 \times 10^{10} \, h^{-2} \, L_\odot \tag{3c}$$

(Kirshner, Oemler & Schechter 1979). The majority of galaxies are faint but most of the light comes from those that are more luminous than 0.7L*. Thus, L* may be taken as the characteristic luminosity of bright galaxies and ϕ* may be taken as an estimate of their average density. By integrating $L\phi(L)$ over L, the cosmic luminosity density and mass-to-light ratio can be derived:

$$L \simeq 1.9 \times 10^{8} hL_{\odot} \ \mathrm{Mpc}^{-3} , \qquad\qquad (4a)$$

$$<M/L> \equiv \rho/L \simeq 1500\Omega h. \qquad\qquad (4b)$$

Here $<M/L>$ is in solar units and Ω is the cosmic mass density ρ in units of the critical density required to close the universe. The internal velocities of groups and clusters typically imply mass-to-light ratios of 150h to 500 h, which corresponds to $\Omega \simeq 0.1$–0.3 unless mass and luminosity are segregated on scales larger than several megaparsecs.

The shapes of the luminosity functions for the main classes of galaxies differ at their faint ends (L $\leq$ 0.1 L*), dwarf E and I galaxies being more numerous than dwarf S0 and S galaxies. At their bright ends, however, the shapes of the luminosity functions for S, S0 and normal E galaxies (excluding the cD type) are similar, with similar values of L* but different amplitudes ϕ* (Tammann, Yahil & Sandage 1979). At the characteristic luminosity L*, the observed values of the effective radii are comparable for these classes:

$$r_{e}^{*} \simeq (2 - 3)h^{-1} \ \mathrm{kpc} \quad (E), \qquad\qquad (5a)$$

$$r_{e}^{*} \simeq (3 - 5)h^{-1} \ \mathrm{kpc} \quad (S \ \& \ S0), \qquad\qquad (5b)$$

where the values in common are mainly for S0 and S galaxies with small disk-to-bulge ratios (de Vaucouleurs, de Vaucouleurs & Corwin 1976). Galaxies with the dimensions given by (5) represent enhancements in luminosity density above the cosmic value (4a) by very significant factors:

$$\delta* \simeq (\tfrac{1}{2}L*)/(\tfrac{4}{3} \pi r_{e}^{*3}L) \sim 10^{8} - 10^{9}. \qquad\qquad (6)$$

From (3c) and (4b), it follows that the characteristic mass associated with bright galaxies is about

$$M* \simeq 1.5 \times 10^{13} \ \Omega h^{-1} \ M_{\odot} . \qquad\qquad (7)$$

Some of this is gravitationally bound to individual galaxies as dark halos but much of it may be in the form of intergalactic material, especially in groups and clusters where the time-scales for tidal stripping are short.

5

5 STELLAR CONTENT

It is often convenient to refer the stellar populations of galaxies to their structural components instead of the traditional categories, populations I and II, etc. In this context, globular clusters are usually associated with bulges, although a distinction between them may be appropriate for some purposes. In the Milky Way, this population has a wide range of metallicity below the solar value, $0.003Z_\odot \lesssim Z \lesssim 0.6Z_\odot$, and a radial gradient in the sense that the average values of Z decrease with increasing galacto-centric distance. Its constituents are roughly $(12-16) \times 10^9$ yrs old, where the indicated age-spread may or may not be real. The disk population of the Milky Way has a narrow range of metallicity near the solar value, $0.3Z_\odot \lesssim Z \lesssim 2Z_\odot$, and comparatively weak gradients in both the radial and vertical directions. The ages of its open clusters range from zero for those now forming in spiral arms up to at least 5×10^9 yr for NGC 188, the oldest known example (Demarque & McClure 1977). This is comparable with the age of the sun and represents only a lower limit to the age of the disk because any open clusters that formed earlier might have been tidally dissolved. Taken as a whole, the current evidence suggests that pregalactic material was metal-poor, that the bulge formed during a phase of rapid enrichment and that the disk formed sometime later.

From their composite spectra and colours, most galaxies are found to have metallicity gradients and central metallicities that correlate with total luminosities. Using a recent calibration of the Mg index, the average relations for bright ellipticals can be summarized as follows:

$$d(\log Z)/d(\log r) \simeq -0.3, \tag{8a}$$

$$Z_c/Z_\odot \simeq 2 \ (L/L\ast)^{0.2}, \tag{8b}$$

(Faber 1977). In many respects, the stellar content of E and S0 galaxies appears to be similar to that in the bulge of the Milky Way and therefore to be old; nevertheless, all that is certain is that a negligible fraction of their stars formed within the last 10^9 yr (Larson & Tinsley 1978). By contrast, spiral disks have the earlier spectral types and bluer colours indicative of current star formation. In fact, these change along the spiral sequence in the sense expected from the strength of their arms, with $B-V \simeq 0.7$ at Sa and $B-V \simeq 0.4$ at Sd. Star formation is especially vigorous in some I galaxies, intergalactic HII regions and interacting galaxies. Certain environments may be responsible for even more dramatic changes in the stellar content of galaxies, as suggested by a study of several galaxy clusters with redshifts up to 0.4 and therefore look-back times up to $4 \times 10^9 \ h^{-1}$ yr (Butcher & Oemler 1978). In the distant clusters, the proportion of galaxies with the colours of spirals appears to be about three times higher than

it is in nearby clusters of comparable richness, possibly because
spirals are being converted into lenticulars.

6 GAS CONTENT

Four principal components or 'phases' of the interstellar
medium have been identified in the Milky Way. They appear to be
roughly in pressure balance with each other and to exchange mass
and energy in a feedback cycle regulated by the births and deaths
of stars. The ionized gas has a hot component $(T \sim 10^6)$, which is
extremely diffuse, and a warm component $(T \sim 10^4 K)$, which has a
patchy distribution. Most of the gas, however, is cool enough to
be neutral, in either atomic HI $(T \sim 10^2 K)$ or molecular H_2 $(T \sim 10K)$
forms. These components are confined to the disk, having scale
heights of 120 pc (HI) and 60 pc (H_2) near the sun, and their distributions are distinctly clumpy on such scales. Typical masses of
HI cloud complexes and giant molecular clouds are of order $(10^5 -
10^6)M_\odot$; the total mass of the H_2 component is uncertain but it may
be as high as 3×10^9 $M_\odot$ and therefore comparable with the mass of
the HI component (Scoville & Solomon 1975). The dust-laden molecular clouds are generally associated with spiral features and are
known to be conducive to star formation. Once this begins, the
clouds are thought to develop into HII regions, ionized by the
ultraviolet flux of newly-formed O and B stars, and then into
young clusters as the gas is depleted or expelled. Unfortunately,
next to nothing is known about the molecular hydrogen content of
external galaxies.

The distribution of atomic hydrogen has been mapped at high
resolution in many galaxies (van der Kruit & Allen 1978). On
average, the ratio of mass in HI to optical luminosity increases
steadily through the spiral sequence, from $M_H/L \simeq 0.07$ at Sa to
$M_H/L \simeq 0.5$ at Sd. The gas often extends beyond the Holmberg radius
and is often depleted in the bulge regions of the disk. In most
cases, the random velocities of the gas are about 10 km s^{-1} over
the entire disk and the density of the gas is 2 to 10 times higher
in the arm regions than it is in the interarm regions. The HI
layers are thinner than the stellar disks and usually have scale
heights that increase gradually out to r_H; beyond r_H they are often
flared or warped. I galaxies are generally gas-rich, whereas SO
galaxies have a surprisingly wide range of gas content, with
$0.01 \leq M_H/L \leq 0.5$. By comparison, E galaxies are generally gas-
poor, the limits being $M_H/L \leq 0.03$ in most cases. The expected
rate of mass-loss from an old stellar population is of order
$(L/10^{11}L_\odot)M_\odot$ yr^{-1}, enough to provide bright ellipticals and bulges
with a detectable quantity of gas in about 10^9 yr, so this must
somehow have been removed. It is likely that at least part of the
effect can be explained on the basis of hot winds $(T \sim 10^6 K)$ driven
by supernova explosions (Mathews & Baker 1971). Intergalactic HI
clouds have been found in the vicinity of interacting galaxies and
in some groups but truly isolated clouds seem to be rare.

7 INTERNAL KINEMATICS

Disk stars in the solar neighbourhood ($r_\bullet \simeq 9$ kpc) have average rotation velocities near the local circular velocity, $v_\bullet \simeq 200\text{-}230$ km s^{-1}. Their velocity dispersions along cylindrical coordinate axes range from $\sigma_r \simeq \sigma_\phi \simeq \sigma_z \simeq 10$ km s^{-1} for the young O and B stars up to $\sigma_r \simeq 50$, $\sigma_\phi \simeq 30$, $\sigma_z \simeq 25$ km s^{-1} for the oldest K and M stars. These velocity ellipsoids are nicely explained, both in volume and shape, by the diffusion of orbits through encounters with irregularities in the disk, giant molecular clouds being the most likely agents. This process has an age-dependence of the form

$$\sigma_k^{\,2}(\tau) = \sigma_g^{\,2} + C_k \tau \qquad (k = r, \phi, z), \qquad (9a)$$

$$C_r = 1.75D, \quad C_\phi = 0.70D, \quad C_z = 0.50D, \qquad (9b)$$

where the initial dispersion, $\sigma_g \simeq 10$ km s^{-1}, is that in the star-forming gas and the required diffusion coefficient, $D \simeq 2 \times 10^{-7}$ (km s^{-1})^2 yr^{-1}, implies perturbing masses of order 10^6 $M_\bullet$ (Wielen 1977). From the z-velocities of nearby stars, the local mass-to-light ratio in the disk is estimated to be about 3, most of which is accounted for by gas and faint stars (Oort 1960). The metal-poor RR Lyrae variables are probably the most reliable tracers of the bulge; in the solar neighbourhood, they have $\sigma_r \simeq 140$, $\sigma_\phi \simeq 120$, $\sigma_z \simeq 75$ km s^{-1} (Woolley 1978), but little is known about their velocity ellipsoid in the central regions of the Milky Way. The system of globular clusters has an average rotation velocity of $\bar{v} = 60 \pm 25$ km s^{-1} and a velocity ellipsoid with $\sigma_r \simeq \sigma_\phi \simeq \sigma_z \simeq 120$ km s^{-1}; hence $\bar{v}/\sigma \simeq 0.5$ (Frenk & White 1980).

The rotation curves of most disk galaxies rise almost linearly out to some 'turnover' radius r_m, which is usually smaller than twice the exponential scale length α^{-1}. Beyond this, they are roughly constant at some velocity v_m out to the limit of detection for 21cm and emission lines, which is usually comparable with the Holmberg radius r_H. These flat rotation curves imply the existence of dark halos and, if they are approximately spherical, the required density profiles are of isothermal form, with $\rho(s) \simeq v_m^2/4\pi G s^2$ and $s^2 = r^2 + z^2$ in the outer parts ($s \gtrsim 2\alpha^{-1}$). The enclosed mass is $M(s) \simeq s v_m^2/G$ or, equivalently,

$$M(s) \simeq 2.2 \times 10^{11} (s/r_H)(L_D/L*)^{\frac{1}{2}}(v_m/250)^2 h^{-1} M_\bullet, \qquad (10)$$

for an exponential disk with $I_0 = 145$ $L_\bullet pc^{-2}$. Galaxies with luminosities near L* usually have rotation velocities in the range 200 km s^{-1} $\leq v_m \leq 300$ km s^{-1} and, when the luminosities of their bulges are included, the estimated mass-to-light ratios at r_H are usually in the range $5h \leq M(r_H)/L \leq 20h$ (Bosma 1978, Rubin, Ford & Thonnard 1978). Most of this spread reflects different contributions by young stars to the B-luminosities of galaxies with different morphological types, Sa galaxies having higher values of $M(r_H)/L$ than Sd galaxies. In the infrared, however, luminosities and velocities are found to correlate as $v_m \propto L^{\frac{1}{4}}$ (Aaronson, Huchra & Mould 1979), and a relation similar to (10) then implies that $M(r_H)/L$ is nearly constant at a value about twice that inferred for the visible components. Candidates for the 'missing' mass include faint stars, stellar remnants and massive neutrinos among many other possibilities.

Most E galaxies have rotation velocities $\bar{v}$ and velocity dispersions σ along the line of sight that are roughly constant out to r_e, the point at which absorption-line measurements become difficult. From the distribution of velocity ratios $(\bar{v}/\sigma)$ over axial ratios (b/a), it follows that they have anisotropic velocity ellipsoids and that rotation plays only a minor role in their support, the average velocity ratio being $<\bar{v}/\sigma> \simeq 0.3$ (Binney 1978). From the virial theorem in some form or other, the central mass-to-light ratios of bright E galaxies are estimated to be about $(10-20)h$, but little is known about the distribution of mass in their outer parts. On average, their central velocity dispersions correlate with total luminosities as

$$\sigma_c \simeq 220 \ (L/L*)^{\frac{1}{4}} \ \mathrm{km \ s}^{-1} \tag{11}$$

(Faber & Jackson 1976). Comparing this with (8b) indicates that metallicities are nearly proportional to velocity dispersions and therefore to escape velocities. In fact, even the deviations from these average relations are correlated in the sense that higher metallicities are usually associated with higher dispersions at a fixed luminosity (Terlevich et al. 1980). This means that E galaxies lie on a surface in the $(L, \bar{Z}_c, \sigma_c)$ space and that two parameters, in addition to $\bar{v}/\sigma$ and perhaps b/a, are needed to specify their properties. In some respects, the bulges of S0 and S galaxies are similar; however, the dozen or so with measured values of $\bar{v}$ and σ appear to require velocity ellipsoids that are more nearly iso-tropic than those of bright E galaxies (Illingworth & Kormendy 1980).

8 GROUPS AND CLUSTERS

Most galaxies are members of some aggregate in the range between poor groups and rich clusters. The best-studied example of a poor group is the Local Group, which has a total luminosity of about $4h^2L*$ and consists of the Milky Way, Andromeda (M31) and about twenty dwarf galaxies. It has a velocity dispersion of about 60 km s^{-1} and an effective radius of about 0.7 Mpc so the inferred mass is about 5×10^{12} M$_\odot$. The best studied example of a rich cluster is Coma (Abell 1656), which has a total luminosity of about 600 L* and consists mainly of E and S0 galaxies. It has a velocity dispersion of about 900 km s^{-1} and an effective radius of about $2.0h^{-1}$ Mpc so the inferred mass is about $3 \times 10^{15}h^{-1}$M$_\odot$ (Rood et al.

1972). The cluster is significantly flattened, with an apparent axial ratio of about 0.5, but it has no signficant rotation along the line of sight, with an upper limit of $\overline{v}/\sigma \leq 0.2$ over the main body. X-ray emission from the cluster indicates that it contains about $2 \times 10^{14} h^{-3/2} M_\odot$ in the form of ionized gas at a temperature of about 10^8 K. The strength of the 7kev iron line in this emission suggests a metal abundance near the solar value so the gas was probably expelled from member galaxies (Serlemitsos et al. 1977).

An important method for describing the general pattern of clustering is in terms of correlations between the positions of galaxies. The basic element of this description is the pair correlation function, defined as the probability, in excess of the Poisson value, of finding two galaxies separated by a distance r. It has the approximate power-law form

$$\xi(r) = (r_o/r)^\gamma, \tag{12a}$$

$$\gamma \simeq 1.8, \qquad r_o \simeq 4 \, h^{-1} \, \text{Mpc}, \tag{12b}$$

for separations between about $20 \, h^{-1}$ kpc and $10 \, h^{-1}$ Mpc (Davis, Geller & Huchra 1978). A few of the higher-order correlation functions have also been estimated and they are consistent with the notion of a hierarchy in which small aggregates of high density are most often located inside large aggregates of low density (Fry & Peebles 1978). In this case, $\xi(r)$ is a good measure of the typical density enhancement on the scale r. An extrapolation of (12) down to the characteristic scale (5) of bright galaxies gives $\xi(r_e*) \simeq 10^5 - 10^6$, which falls short of their actual density enhancement (6) by factors of order 10^3. Thus, the luminous parts of galaxies are about ten times smaller than they would be if they joined smoothly onto the clustering hierarchy. On scales larger than r_o the deviations from a uniform distribution are generally small and therefore difficult to measure. There is, however, some evidence for chain-like structures that are separated by relatively empty regions with dimensions of order $50 \, h^{-1}$ Mpc.

REFERENCES

Aaronson, M., Huchra, J. & Mould, J. (1979). Astrophys.J., 229, 1.
Binney, J. (1978). Mon.Not.R.Astr.Soc., 183, 501.
Bosma, A. (1978). Ph.D. thesis, University of Groningen.
Burstein, D. (1979). Astrophys.J., 234, 435.
Butcher, H. & Oemler, A. (1978). Astrophys.J., 226, 559.
Davis, M., Geller, M.J. & Huchra, J. (1978). Astrophys.J., 221, 1.
Demarque, P. & McClure, R.D. (1977). In The Evolution of Galaxies and Stellar Populations, eds. B.M. Tinsley & R.B. Larson, Yale University Observatory, p. 199.
Dressler, A. (1980). Astrophys.J., 236, 351.

Faber, S.M. (1977). In The Evolution of Galaxies and Stellar
    Populations, eds. B.M. Tinsley & R.B. Larson, Yale University
    Observatory, p. 157.
Faber, S.M. & Jackson, R.E. (1976). Astrophys.J., 204, 668.
Freeman, K.C. (1970). Astrophys.J., 160, 811.
Frenk, C.S. & White, S.D.M. (1980). Mon.Not.R.Astr.Soc., 193, 295.
Fry, J.N. & Peebles, P.J.E. (1978). Astrophys.J., 221, 19.
Illingworth, G. & Kormendy, J. (1980). In preparation.
Kirshner, R.P., Oemler, A. & Schechter, P.L. (1979). Astron.J.,
    84, 951.
Kormendy, J. (1977). Astrophys.J., 218, 333.
Larson, R.B. & Tinsley, B.M. (1978). Astrophys.J., 219, 46.
Mathews, W.G. & Baker, J.C. (1971). Astrophys.J., 170, 241.
Oort, J.H. (1960). Bull.Astr.Inst.Netherlands, 15, 45.
Rood, H.J., Page, T.L., Kintner, E.C. & King, I.R. (1972).
    Astrophys.J., 175, 627.
Rubin, V.C., Ford, W.K. & Thonnard, N. (1978). Astrophys.J., 225,
    L107.
Scoville, N.Z. & Solomon, P.M. (1975). Astrophys.J.Lett., 199,
    L105.
Serlemitsos, P.J., Smith, B.W., Boldt, E.A., Holt, S.S. & Swank, J.H.
    (1977). Astrophys.J.Lett., 211, L63.
Strom, K.M. & Strom, S.E. (1978). Astron.J., 83, 1293.
Tammann, G.A., Yahil, A. & Sandage, A. (1979). Astrophys.J., 234,
    775.
Terlevich, R., Davies, R.L., Faber, S.M. & Burstein, D. (1980).
    Mon.Not.R.Astr.Soc., in press.
Toomre, A. & Toomre, J. (1972). Astrophys.J., 178, 623.
van der Kruit, P.C. & Allen, R.J. (1978). Ann.Rev.Astr.Astrophys.
    16, 103.
de Vaucouleurs, G. & Pence, W.D. (1978). Astron.J., 83, 1163.
de Vaucouleurs, G., de Vaucouleurs, A. & Corwin, H.G. (1976).
    Second Reference Catalogue of Bright Galaxies, University of
    Texas Press, Austin.
Wielen, R. (1977). Astr.Astrophys., 60, 263.
Woolley, R. (1978). Mon.Not.R.Astr.Soc., 184, 311.
Yoshizawa, M. & Wakamatsu, K. (1975). Astr.Astrophys., 44, 363.

PHOTOMETRIC AND DYNAMICAL PROPERTIES
OF ELLIPTICAL GALAXIES

Francesco Bertola

Istituto di Astronomia, Padova

1. INTRODUCTION

Although a more complicated picture emerges from recent developments, normal elliptical galaxies can be considered as relatively
simple systems. They are thought to be one-component stellar aggregates, whereas the remaining galaxies are composite systems. At
least two components, bulges and disks, are recognized in SO and
spiral galaxies and they play a significant role in morphological
descriptions and theories of galaxy formation.

From the morphological point of view, elliptical galaxies
resemble somewhat the bulges of SO and spiral galaxies. It seems,
however, that the bulge components of disk galaxies have kinematic
(Bertola & Capaccioli 1977; Kormendy & Illingworth 1979) and colour
properties (Strom & Strom 1978) that distinguish them from ellipticals. Elliptical and SO galaxies are often treated together in the
literature under the term 'early type galaxies'. Although this
grouping can be practical for some purposes, the mixing of E and
SO types can also cause misunderstandings. A significant example
of this has occurred in the treatment of the differential velocities
of E and SO pairs (Faber & Gallagher 1979). The distribution of Δv
for E pairs suggests that they may not be physically associated,
whereas the distribution of Δv for SO galaxies is typical of physical
pairs. This fact was hidden for some time by samples that combined
the two types.

A number of properties indicate that there is a true dichotomy
between these galaxies and that the SO type should not be regarded
as a transition stage between the elliptical and spiral types (Bertola
& Capaccioli 1978). Instead, it appears to form a sequence parallel
to the spiral sequence of increasing disk-to-bulge ratio, as in van
den Bergh's (1976) classification scheme. The presence of a disk
seems to be the discriminating factor.

13

F. BERTOLA

2. PHOTOMETRIC PROPERTIES

2.1 Luminosity Profiles

Several laws have been proposed to represent the observed
luminosity profiles of elliptical galaxies (Hubble 1930; de
Vaucouleurs 1953; Baum 1955; King 1966; Oemler 1976). The most
widely used are those of de Vaucouleurs and King. de Vaucouleurs'
law is strictly empirical and is given by the formula

$$\log (I/I_o) = -3.33071 (r/r_e)^{1/4}.$$

Here I_o and r_e are scale factors, which, for an ideal EO represent
the peak brightness and the radius containing half the total light.
This law has only two scale factors and no free parameters; it is
easily compared with observations since it is a straight line in
the $\log I$-$r^{1/4}$ plane.

King's law is semi-empirical and was derived from quasi-
isothermal models to fit the projected distribution of stars in
globular clusters. It can also be used to represent the luminosity
distributions in elliptical galaxies by the formula

$$I = K \left[(1 + r^2/r_c^2)^{-\frac{1}{2}} - (1 + r_t^2/r_c^2)^{-\frac{1}{2}} \right]^2.$$

Of the three constants in this expression, two have the role of
scale factors, say K and r_c, and the third r_t is a free parameter.
The quantity r_c is referred to as the 'core radius'; it is the
distance from the centre, in a direction 45° from the principal
axes, at which the surface brightness is approximately half the
central value. The quantity r_t is referred to as the 'tidal radius',
and it is the distance from the centre where the surface brightness
drops to zero. Typical values of r_t/r_c are 100 to 200 for giant
ellipticals. King's formula often fails to fit the outer parts of
these galaxies and it is sometimes difficult to estimate r_c as the
result of seeing effects (Schweizer 1979).

Recent work indicates that most elliptical galaxies follow the
$r^{1/4}$ law rather closely. Thus, an observed profile can be conveniently described by the departures from this law, which are
usually most pronounced in the innermost and outermost regions.
By way of illustration, we consider the standard galaxies M87 and
NGC 3379. Faint envelopes have been detected in the outer parts
of M87 and several other ellipticals (Arp & Bertola 1969, 1971;
Oemler 1976). This phenomenon may depend on whether a galaxy is
isolated or has close neighbours (Strom & Strom 1978; Kormendy 1977).
Concentrations of light have been detected in the inner parts of
both M87 (Young et al. 1978a) and NGC 3379 (de Vaucouleurs & Nieto and
Cappacioli 1979).

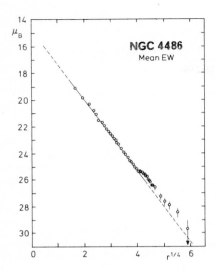

Fig. 1. The luminosity profile of M87 in the outer
regions where the envelope
appears (from de Vaucouleurs
& Nieto 1978).

The photometry of M87 shown in Figure 1 has a departure from
the $r^{1/4}$ law in the form of a hump (de Vaucouleurs & Nieto 1978).
It begins at r = 4!5 on the east-west profile, reaches a maximum
at 5' < r < 6' and is followed by an exponential decline out to
r = 20', where the surface brightness is 29 mag/sq arcsec or 1/500
of the night sky. This envelope contributes about 8% or less to
the total light of the galaxy. Arp & Bertola (1971) found that a
sample of ellipticals in Perseus, covering a wide range in luminosities, had similar features in their outer parts. Most of the
region of M87 interior to r = 18" is fainter than the extrapolated
$r^{1/4}$ law, as shown in Figure 2. Within 1" of the centre, there is
a central spike that contributes about 11% of the light interior
to r = 6" (Young et al. 1978a).

The departure from the $r^{1/4}$ law in NGC 3379 is in the form of
an excess of light near the centre. After deconvolution for seeing,
it can be fitted by a Gaussian profile superimposed on the $r^{1/4}$
component. The extra light in the central component is about 4%
of the total luminosity. As Figure 3 shows, the outer profile of
NGC 3379 follows the $r^{1/4}$ law to the last measured point at r = 7!3,
where the surface brightness is 27.8 mag/sq arcsec. Unlike M87,
this nearby galaxy has little or no extra light in its outer parts.

2.2 Ellipticity Changes and the Isophote Twists

Only recently has attention been drawn to radial changes and
twisting of the isophotal contours in elliptical galaxies. These

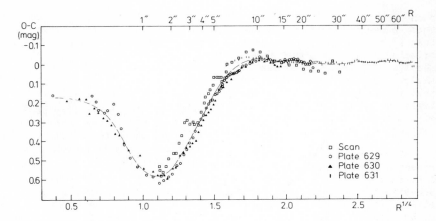

Fig. 2. Deviations from the $r^{1/4}$ law in the inner parts of M87 showing a general deficit of light and part of the central spike. The data have been deconvolved with a point spread function to correct for seeing (from de Vaucouleurs & Nieto 1979).

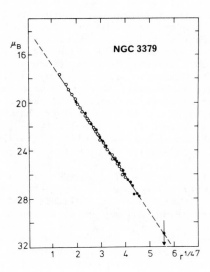

Fig. 3. The luminosity profile of NGC 3379 in the outer regions where the $r^{1/4}$ law is followed accurately (from de Vaucouleurs & Capaccioli 1979).

phenomena are important because they contain information about the three-dimensional shapes of the objects. It can easily be shown that if the surfaces of constant luminosity density are non-similar, triaxial ellipsoids, then the observed isophotes will have radial changes of ellipticity and twisting of the axes at most viewing angles. As discussed below, recent measurements of the internal velocities of elliptical galaxies indicate that they have anisotropic velocity distributions. A variety of theoretical arguments suggests that such systems can have oblate or prolate forms in the simplest cases, and triaxial forms in the most general case.

Both radial changes of ellipticity and twisting of axes have recently been reported by a number of authors (Carter 1978; Strom & Strom 1978; King 1978; Bertola & Galletta 1979; Williams & Schwarzschild 1979). All possible trends appear to be present; ellipticities may increase, decrease, have maxima, have minima or be constant with distance from the centre of a galaxy. The variations in the position angles of the axes may be gradual or they may be abrupt. Figure 4 shows the isophotes of NGC 4125, an E5 galaxy with both radial changes of ellipticity and twisting of axes.

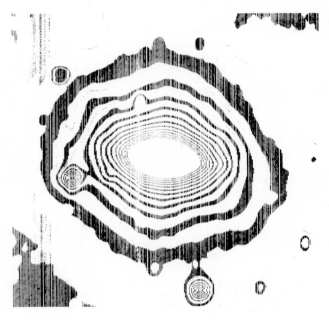

Fig. 4. Isophotal map of NGC 4125 from a deep IIIa-J plate taken with the 48-inch Palomar Schmidt. The scale is 4.9"/mm and the contours are at increments of 0.04 in photographic density.

3. DYNAMICAL PROPERTIES

The kinematics of elliptical galaxies can be described by the observed rotation velocities and velocity dispersions in their projected images; they are the result of integration along the line of sight. Early observations gave only the central velocity dispersions of elliptical galaxies. The first rotation curve, that of the E5 galaxy NGC 4697, was produced in 1972 (Bertola 1972; Bertola & Capaccioli 1975). Since then a large amount of data has been collected. Rotation curves along different axes have been measured for about 30 galaxies and central velocity dispersions have been measured for about 50 galaxies (Capaccioli 1979). In most cases, profiles of velocity dispersion are also available.

3.1 Rotation Curves

The unexpected property of NGC 4697 was its low rotation velocity, amounting to only 65 km s^{-1}, despite an appreciable flattening of the galaxy. Subsequent to this discovery, slow rotation has been found in most of the ellipticals for which velocity data have been taken; in almost all cases, the maximum rotation velocity is less than 100 km s^{-1}. These observations caused the abandonment of models in which flattening is entirely the result of rotation (Illingworth 1977). Among the models with rapid rotation are the self-consistent ones developed by Prendergast & Tomer (1970) and Wilson (1975), the dissipationless collapse models of Gott (1975) and the dissipative infall models of Larson (1975). The failure of these models to represent real elliptical galaxies means that their velocity distributions are anisotropic (Binney 1980). This fact has stimulated much recent work on the structure and dynamics of elliptical galaxies.

Rotation has been searched for along the minor axis as well as the major axis of several ellipticals. Usually, there is little or no evidence for minor axis rotation but it has been detected in at least two cases: NGC 596 (Schechter & Gunn 1979) and NGC 4125 (Bertola & Capaccioli 1980). This phenomenon is important because it is one of the crucial tests for triaxiality.

3.2 Velocity Dispersions

A measure of the random motions of stars in a galaxy, the velocity dispersion, is obtained from the width of absorption lines in its spectrum. Precise measurements using modern detectors and reduction techniques have led to the following mean relation between the luminosity and the central velocity dispersion of ellipticals: $L \propto \sigma^4$ (Faber & Jackson 1976).

It has recently been shown that the scatter in this relation and the analogous one relating luminosity and metallicity can be reduced if corrections are applied to the measured line widths and line strengths (Terlevich et al. 1980). These two corrections, which are highly correlated, are in turn correlated with apparent axial ratio. This suggests that elliptical galaxies are character- ised by a second fundamental parameter in addition to luminosity. This may be related to their intrinsic axial ratios and intrinsic shapes, i.e. whether they are oblate, prolate or triaxial. The second parameter is almost certainly an important clue to the way in which elliptical galaxies formed.

3.3 Masses and Mass-to-Light Ratios

Because elliptical galaxies are not primarily supported by rotation, their rotation curves are of little help in determining masses. Instead, the central velocity dispersion, or dispersion profile when available, is of greater utility. The fact that few elliptical pairs are physically associated means, unfortunately, that the method of binary galaxies cannot be used to determine their masses (Faber & Gallagher 1979). Three methods have been used to derive the masses and mass-to-light ratios of ellipticals.

The first method, due to Poveda (1958), is based on the virial theorem applied to the entire galaxy: $2T + \Omega = 0$. The total kin- etic energy, $T = \frac{1}{2}M\sigma^2$, is usually estimated from the central velo- city dispersion, assumed to be characteristic of the whole. The total potential energy Ω is computed from the $r^{1/4}$ profile on the assumption that the mass-to-light ratio is constant throughout the galaxy. For a spherical system, the derived mass is

$$M = 0.2 \ r_e d\sigma^2$$

where r_e is in arcmin, σ is in km s^{-1} and d, the distance of the galaxy, is in parsec. A simple correction can be applied to this formula if the galaxy is flattened.

The dubious assumptions of the previous method can be overcome by estimating a local mass-to-light ratio, usually in the core regions. For a King model, which has an isotropic, Gaussian distribution of velocities in the core, one finds

$$\rho_o = 9\sigma_o^2/4\pi Gr_c^2$$

for the central mass density (Rood et al. 1972). In this expression, the true velocity dispersion in one dimension σ_o can safely be replaced by the measured value at the centre of the projected image. If the central surface brightness, properly corrected for seeing effects, is known, then the mass-to-light ratio of the core region is easily derived. By this method, Faber & Jackson (1976) and Davies (1980) estimated $M/L_B \simeq 8$ for a sample of 16 giant ellip- ticals.

Profiles of velocity dispersion give further dynamical infor- mation as shown in Figure 5. With such data, the mass density of a galaxy can be estimated as a function of distance from its centre using the equations of stellar hydrodynamics. The first-order moment equation for a spherical system is

$$\frac{1}{\rho} \frac{d}{dr} (\rho\sigma_r^2) + \frac{2}{r} (\sigma_r^2 - \sigma_t^2) = \frac{-GM(r)}{r^2}$$

where ρ is the density of luminous stars, σ_r and σ_t are, respectively, the radial and tangential velocity dispersions of these stars and M(r) is the total mass enclosed within the radius r. This formula has been used by Efstathiou, Ellis & Carter (1980) to estimate the masses and mass-to-light ratios of the 13 ellipticals observed by them and Schechter & Gunn (1979). They find masses of order $10^{11} M_\odot$ and $M/L_B \simeq 7$ at $<r_{max}> \simeq 5$ kpc, in good agreement with estimates for the core regions.

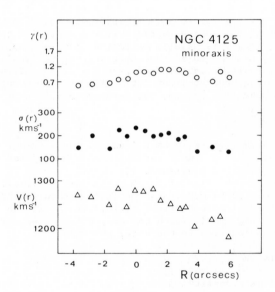

Fig. 5. Spectral data for the minor axis of NGC 4125, taken with the 4m Kitt Peak telescope and reduced by the Fourier quotient method (Bertola et al. 1980). The quantity γ is the average absorption line strength ratio between the galaxy and a template star. Notice the tendency for minor axis rotation and an outward decrease in velocity dispersion. This galaxy also appears to have some gas with complicated dynamics (Bertola & Capaccioli 1980).

ELLIPTICAL GALAXIES

Velocity dispersion profiles are especially interesting in the central regions of elliptical galaxies; examples are shown in Figures 6 and 7. In M87, σ has been found to rise toward the centre and this has been interpreted as evidence for a mass concentration, perhaps a black hole, of 5 x 10^9 $M_\odot$ within the central 100 pc (Young et al. 1978a; Sargent et al. 1978). As Duncan & Wheeler (1980) have pointed out, it can also be interpreted as evidence for velocity anisotropy, without such a large mass concentration. Recently, Dressler (1980) has found that σ levels off within 1" of the centre of M87, contrary to the results of Sargent et al. (1978); he suggests that this and the luminous spike can be explained in terms of a dense cluster of stars. A rise in the velocity dispersion has also been found near the centre of NGC 3379, which is generally regarded as a normal elliptical.

In the outer regions of elliptical galaxies, the run of velocity dispersion may provide clues about whether they have massive halos of dark material. The velocity dispersion in Davies' (1980) sample has a tendency to decrease at large radii, suggesting the absence of such halos. On the other hand, the measurements by Young et al. (1978b), Schechter & Gunn (1979) and Efstathiou et al. (1980) have nearly constant velocity dispersion profiles. This would imply, under certain assumptions, that the mass distribution is of the form $M(r) \propto r$, as indicated by the flat rotation curves of spiral galaxies (Efstathiou et al. 1980). In the cD galaxy of Abell 2029, the velocity dispersion increases outward (Dressler 1979) whereas, in the cD galaxy of Abell 401, the velocity dispersion is the same in the inner and outer parts (Faber, Burstein & Dressler 1977). Both of these results imply that M/L is higher in the envelopes than in the central regions.

A new approach to the determination of the masses of elliptical galaxies makes use of X-ray data. The equation of hydrostatic equilibrium for an ideal gas gives the following mass-radius relation:

$$M(r) \propto rT \, d \log \rho / d \log r$$

where ρ is the density of the gas and T is its temperature, which is assumed to be constant in this example. The logarithmic density gradient can be inferred from the observed profile of X-ray surface brightness and the temperature can be inferred from the energy distribution of X-ray counts. In this way, Fabricant, Lecar & Gorenstein (1980) have derived the relation $M(r) \simeq 1.2$ x $10^{13} M_\odot$ (r/100 kpc) for M87. Their data extend to r = 230 kpc and give a total mass much greater than the optical data, which refer only to the luminous parts of the galaxy. The conclusion therefore is that a massive halo of dark material surrounds M87 in addition to the previously discovered optical envelope (Arp & Bertola 1969).

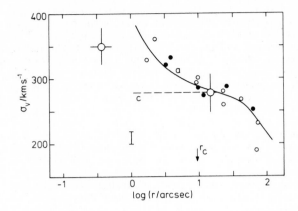

Fig. 6. Velocity dispersion profile near the centre of M87. Full and open circles and the solid line are from Sargent et al. (1978); the crossed circles are from Dressler (1980).

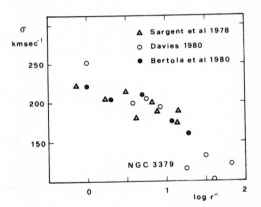

Fig. 7. Velocity dispersion profile near the centre of NGC 3379 from three sets of measurements: Sargent et al. (1978), IPCS on the 5m Palomar telescope; Davies (1980), IPCS on the 3.9m Anglo-Australian telescope; Bertola et al. (1980), SIT Vidicon on the 5m Palomar telescope. Notice the increase of σ to the centre, as in M87, and the good agreement between observers.

4. THREE-DIMENSIONAL SHAPES

The idea, advocated by Binney (1978), that elliptical galaxies are generally triaxial has recently stimulated much interest in the problem of determining their three-dimensional forms. On the theoretical side, Miller (1978) has produced N-body models of prolate systems and Schwarzschild (1979) has demonstrated the existence of long-lived triaxial systems in dynamical equilibrium. On the observational side, several new tests have been proposed to determine intrinsic shapes.

Marchant & Olson (1979) and Richstone (1979) have studied the relation between the surface brightness and the flattening of the projected images of a large number of ellipticals. Assuming that intrinsic shape and surface brightness are not correlated, they concluded that there is a preference for oblate over prolate forms. However, the analysis by Binggeli (1980) of the observed distribution of flattenings led him to conclude that the data are consistent with all hypotheses concerning intrinsic shapes. Similar conclusions were reached by Noerdlinger (1979).

The correlations of velocity dispersion and line strength with projected axial ratio, found by Terlevich et al. (1980), suggest several alternative ways of determining the true shapes of elliptical galaxies. The application of these tests to real data, however, has not yet given definitive answers because the presently available sample is too small. Using a triaxial model with variable axial ratios, Benacchio & Galletta (1980) have reproduced the distribution of flattening versus isophotal twisting found by Strom & Strom (1978).

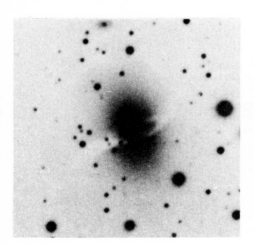

Fig. 8. The elliptical-like galaxy NGC 5266. This object was selected from a list by T.J. Harwarden and photographed for the author by D. Malin with the 3.9m Anglo-Australian telescope; scale 1".5/mm. The orientation of the dust lane may indicate that the galaxy is prolate.

F. BERTOLA

Yet another approach has been suggested by Bertola & Galletta
(1980). It is based on the idea that a disk of gas or dust should
lie in a plane containing the two most similar axes. The prototype
for this kind of test is NGC 5128, which closely resembles an
elliptical except that it has a dust lane crossing the minor axis
of the galaxy. If the dust lane is the projection of a disk seen
edge-on, then one may conclude that the form of this galaxy is
prolate. Figure 8 shows another example of this class of peculiar
galaxies.

ACKNOWLEDGEMENTS

I thank Dr Michael Fall for making many improvements in the
text of this article. This work has been supported in part by the
Gruppo Nazionale di Astronomia, C.N.R.

REFERENCES

Arp, H. & Bertola, F. (1969). Astrophys.Letters, 4, 23.
Arp, H. & Bertola, F. (1971). Astrophys.J., 163, 195.
Baum, W.A. (1955). Publ.Astr.Soc.Pacific, 67, 328.
Benacchio, L. & Galletta, G. (1980). Mon.Not.R.Astr.Soc., in press.
Bertola, F. (1972). Proc. 15th Meeting of the Italian Astronomical
    Society, p. 199.
Bertola, F., Bettoni, D., Rusconi, L. & Sedmak, G. (1980). In
    preparation.
Bertola, F. & Capaccioli, M. (1975). Astrophys.J., 200, 439.
Bertola, F. & Capaccioli, M. (1977). Astrophys.J., 211, 697.
Bertola, F. & Capaccioli, M. (1978). Astrophys.J.Lett., 219, L95.
Bertola, F. & Capaccioli, M. (1980). In preparation.
Bertola, F., Capaccioli, M. & Rose, J. (1980). In preparation.
Bertola, F. & Galletta, G. (1978). Astrophys.J.Lett., 226, L115.
Bertola, F. & Galletta, G. (1979). Astron.Astrophys., 77, 363.
Binggeli, B. (1980). Astron.Astrophys., 82, 289.
Binney, J. (1978). Mon.Not.R.Astr.Soc., 183, 779.
Binney, J. (1980). Phil.Trans.Roy.Soc., 296, 329.
Capaccioli, M. (1979). Photometry, Kinematics and Dynamics of
    Galaxies, ed. D.S. Evans, Univ. of Texas Press, Austin, p.165.
Carter, D. (1978). Mon.Not.R.Astr.Soc., 182, 797.
Davies, R.L. (1980). Mon.Not.R.Astr.Soc., in press.
de Vaucouleurs, G. (1953). Mon.Not.R.Astr.Soc., 113, 134.
de Vaucouleurs, G. & Capaccioli, M. (1979). Astrophys.J.Suppl.,
    40, 699.
de Vaucouleurs, G. & Nieto, J.L. (1978). Astrophys.J., 220, 449.
de Vaucouleurs, G. & Nieto, J.L. (1979). Astrophys.J., 230, 697.
Dressler, A. (1979). Astrophys.J., 231, 659.
Dressler, A. (1980). Astrophys.J.Lett., 240, L11.
Duncan, M.J. & Wheeler, J.C. (1980). Astrophys.J.Lett., 237, L27.

Efstathiou, G., Ellis, R.S. & Carter, D. (1980). Mon.Not.R.Astr. Soc., in press.

Faber, S.M., Burstein, D. & Dressler, A. (1977). Astr.J., 82, 941.

Faber, S.M. & Gallagher, J.S. (1979). Ann.Rev.Astron.Astrophys., 17, 135.

Faber, S.M. & Jackson, R.E. (1976). Astrophys.J., 204, 668.

Fabricant, D., Lecar, M. & Gorenstein, P. (1980). Preprint.

Gott, J.R. (1975). Astrophys.J., 201, 296.

Hubble, E.P. (1930). Astrophys.J., 71, 231.

Illingworth, G. (1977). Astrophys.J.Lett., 218, L43.

King, I.R. (1966). Astron.J., 71, 64.

King, I.R. (1978). Astrophys.J., 222, 1.

Kormendy, J. (1977). Astrophys.J., 218, 333.

Kormendy, J. (1980). Proc. ESO Workshop on Two-Dimensional Photometry, ed. P. Crane & K. Kjär, ESO, Geneva, p. 191.

Kormendy, J. & Illingworth, G. (1979). Photometry, Kinematics and Dynamics of Galaxies, ed. D.S. Evans, Univ. of Texas Press, Austin, p. 195.

Larson, R.B. (1975). Mon.Not.R.Astr.Soc., 173, 671.

Marchant, A.B. & Olson, D.W. (1979). Astrophys.J.Lett., 230, L157.

Miller, R.H. (1978). Astrophys.J., 223, 122.

Noerdlinger, P.D. (1979). Astrophys.J., 234, 802.

Oemler, A. (1976). Astrophys.J., 209, 693.

Poveda, A. (1958). Bull.Obs.Tonantzintla y Tacubaya, 17, 3.

Prendergast, K.H. & Tomer, E. (1970). Astr.J., 75, 674.

Richstone, D.O. (1979). Astrophys.J., 234, 825.

Rood, H.J., Page, T.L., Kintner, E.C. & King, I.R. (1972). Astrophys.J., 175, 627.

Sargent, W.L.W., Young, P.J., Boksenberg, A., Shortridge, K., Lynds, C.R. & Hartwick, F.D.A. (1978). Astrophys.J., 221, 731.

Schechter, P. & Gunn, J. (1979). Astrophys.J., 229, 472.

Strom, S.E. & Strom, K.M. (1978). Astr.J., 83, 73; 732; 1293.

Schwarzschild, M. (1979). Astrophys.J., 232, 236.

Schweizer, F. (1979). Astrophys.J., 233, 23.

Terlevich, R., Davies, R.L., Faber, S.M. & Burstein, D. (1980). Mon.Not.R.Astr.Soc., in press.

van den Bergh, S. (1976). Astrophys.J., 206, 883.

Williams, T.B. & Schwarzschild, M. (1979). Astrophys.J., 227, 56.

Wilson, C.P. (1975). Astr.J., 80, 175.

Young, P., Westphal, J.A., Kristian, J., Wilson, C.P. & Landauer, F.P. (1978a). Astrophys.J., 221, 721.

Young, P., Sargent, W.L.W., Boksenberg, A., Lynds, C.R. & Hartwick, F.D.A. (1978b). Astrophys.J., 222, 450.

GALAXY DYNAMICS: OBSERVATIONS

Garth Illingworth

Kitt Peak National Observatory*
Tucson, Arizona

We have recently seen a remarkable increase in the number of galaxies for which stellar absorption line data have been analyzed to give rotational velocity and velocity dispersion profiles out to large radii — as much as 6-8 kpc in some cases. This is primarily the result of using digital detectors (e.g., the AAO IPCS or the Hale and KPNO SIT systems), or carefully calibrated and processed photographic plates, the data from which have been analyzed with objective and efficient techniques (e.g., see Schechter & Gunn, 1979; Davies, 1980; Fried & Illingworth, 1981; Kormendy & Illingworth, 1981).

At the same time, we have seen that the accuracy to which we can measure velocity dispersions, in particular, has changed significantly from around $\pm25\%$ in the early 1970s to less than $\pm10\%$ now (e.g., see Schechter, 1980; Terlevich et al., 1980; Fig. 2 below). This is in large part due to the widespread use of objective techniques such as the Fourier quotient method developed by Paul Schechter and described in Sargent et al. (1977). This striking improvement, combined with the large radial extent of these new measurements, has revolutionized dynamical studies of galaxies.

Ultimately the real value of these extensive data lies in what can be learned about the era of galaxy formation, the understanding of which must be one of the major goals of astronomy. While the whole process of galaxy formation was very complex and probably rather violent, it is quite likely that galaxies currently retain considerable information about conditions during their formation, as has been emphasized by several authors (e.g., Binney, 1978a, 1979 and other papers in that volume). Clearly then, in any investigation of the formation era one of the first steps is to try to understand the current dynamical state of galaxies.

---

*Operated by the Association of Universities for Research in Astronomy, Inc., under contract with the National Science Foundation.

Interwoven with this, however, is the whole question of the extent and form of the dark material associated with galaxies, since the existence of what may be as much as 90% of the mass of galaxies can only be inferred from its effect upon the structure and dynamics of the visible matter. There is no current evidence to suggest that the dark matter can be directly observed or that its formation affected the remaining small fraction of material in any but minor ways — all in all a rather frustrating situation.

Thus an admirable set of goals for a major observational program on the stellar component in galaxies would appear to be: (1) to define the present dynamical structure of the stellar component of galaxies through spectroscopic data giving rotational velocity and velocity dispersion profiles as well as photometric data giving surface brightness distributions; and (2) to use the above data and additional rotation data to define the existence of the dark material in galaxies as a function of galaxy type, luminosity, and environment, and to delineate the distribution of this material. These data should then be combined with those on the systematic global properties of galaxies throughout the range of Hubble types with the ultimate aim being to establish a consistent picture of the galaxy formation era. The programmes described below attempt to satisfy the spectroscopic aspect of (1) and (2) above. Photometric data are sorely needed.

DARK HALOS ABOUT GALAXIES

Extensive sets of rotation curve data, both in H I (e.g., Bosma, 1978) and in optical emission (e.g., Rubin, Ford & Thonnard, 1980) have been presented over the last few years. The majority of these rotation curves become remarkably flat at large radii, implying that the M/L is increasing strongly with radius (see Bosma and van der Kruit, 1979). These data make a very strong case for the existence of large halos of dark material surrounding spiral galaxies. For early-type disk galaxies, the case has been very much weaker, if it existed at all. However, there are now two edge-on S0 galaxies for which excellent stellar rotation curves exist: NGC 3115 (Rubin, Peterson & Ford, 1980; Illingworth, Schechter & Gunn, 1981) and NGC 4762 (Illingworth & McElroy, 1981). This latter curve is shown in Fig. 1. Both of these galaxies have rotation curves that remain strikingly flat after reaching their maximum velocities. In the case of NGC 4762 the limit is at a radius of nearly 10 (D/21) kpc, where D is the Virgo distance in Mpc.

Similar flat rotation curves have been found for other Virgo S0s in a sample being studied by Mould, Illingworth & Skillman (1981) as part of a project to define the H magnitude—asymptotic velocity relation for S0s, and compare it with that for spirals. These data on NGC 3115, 4762, and the other S0s provide the best evidence for the M/L changes of the form seen in spirals and

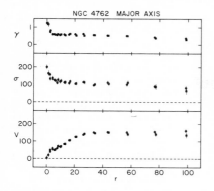

Fig. 1. Line strength (γ), velocity dispersion (σ), and rotation velocity ($V = V_{obs} - V_{systemic}$) data in km s^{-1} folded and plotted against radius in arcsec from a major-axis spectrum of the edge-on S0 galaxy NGC 4762. See Illingworth & McElroy (1981) for details of the interpretation of γ, σ, and V, particularly for $\sigma(r)$ for $r \gtrsim 5''$. Note the flat rotation curve from ~30"-100".

interpreted as being due to the existence of a massive halo. One cannot be too unequivocal here, since the radial extent of such flat rotation curves is very much less in the S0s; however, the data are indicative.

For elliptical galaxies, the evidence is less clear-cut but does indicate, as we shall see below, that M/L changes in the same sense and of similar magnitude do occur (see also Mathews [1978] and Dressler [1979] for discussions on the M/L in M87 and the cD in Abell 2029, respectively). The reason why one is less certain for elliptical galaxies is that the inferred change in M/L arises from velocity dispersion profiles, from which information can be gleaned about only one of the components of the velocity dispersion tensor. Thus the interpretation of the dispersion profiles requires us to make some assumptions about the dynamics of the system. There is no unambiguous measure of the mass distribution like that given by the rotation of a cold component such as gas or disk stars (except possibly in NGC 4278, see Gunn [1979]).

ELLIPTICAL GALAXIES

Since the first indication that elliptical galaxies may not be rotationally flattened (Bertola & Capaccioli, 1975) and the confirmation of this by Illingworth (1977), an extensive body of data has been derived for these systems. Dispersion and rotational velocity profiles along the major axis are now available for 33 galaxies. In addition, minor-axis data are available for some 20% of the sample.

The quality of the data is, in many cases, excellent. A good example is shown in Fig. 2, where data for NGC 4697, taken from several sources, are compared. The agreement between the various velocity dispersion values is good; remarkably so when one considers that three different instruments and data reduction packages were used for the acquisition and analysis of the data

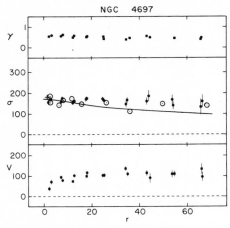

Fig. 2. As for Fig. 1 but with data from major-axis spectra of the elliptical NGC 4697 (Fried & Illingworth, 1981). Additional data from Davies (1980: ⊙) and Schechter (1980: +) are plotted, as is a dispersion profile from an isotropic residual dispersion, constant M/L, log (r_t/r_c) = 2.25 model from King (1966). The model profile is fitted at $r = r_c \approx 2\overset{''}{.}5$ (r_c from King, 1978).

(although in all cases the actual technique used to derive the dispersions was the Fourier quotient method).

As to the data itself there are three noteworthy features: (1) The dispersion profile is essentially flat over radii ranging from $r \approx r_c \sim 2\overset{''}{.}5$ to $r \sim 70''$, nearly 30 r_c (r_c is the core radius) (e.g., see King, 1978). This corresponds to ~7 (D/21) kpc. Such flat dispersion profiles are common in ellipticals (see below). (2) The rotation curve rises rapidly to a maximum at ~1 kpc, after which the rotation is essentially constant out to ~7 kpc. Minor-axis data for this galaxy show no rotation at the 10 km s^{-1} level over several kpc (Davies & Illingworth, 1981). Rotation curve structure such as this is a common feature of elliptical galaxies, even in those cases where the rotation is small and of minimal dynamical importance (see Fried & Illingworth, 1981). (3) $V/\sigma \sim$ 0.6 for this E4 galaxy, making it one of the fastest rotating ellipticals known (in dimensionless terms), having $(V/\sigma)_{obs}/(V/\sigma)_{oblate} \sim 0.7$ (see the discussion and Fig. 4 below). However, it is possible that we may be measuring the rotation from a faint, slightly inclined disk in NGC 4697 (Davies & Illingworth, 1981), and that the bulk of the material in this galaxy may indeed be rotating rather slowly.

The flat dispersion profile is a rather interesting feature of this galaxy. The implications of this can be seen from comparing these data with a velocity dispersion profile from a log r_t/r_c = 2.25 King model. Such a model shows a good match to typical elliptical surface brightness distributions and indicates the expected run of $\sigma(r)$ for a constant M/L, isotropic residual dispersion model. Clearly the data are inconsistent with such a model. The disagreement becomes worse if we have overestimated the core radius. The simplest explanation, in the light of the M/L changes seen in SO and spiral galaxies, is that the M/L is increasing with radius.

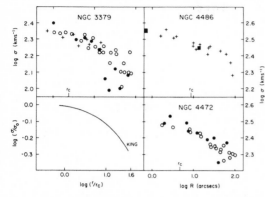

Fig. 3. Dispersion profiles, log σ vs log r, modified from Davies (1980) with data from Sargent et al. (1978: +); Davies (1980: ●); Davies, Illingworth & McElroy (1981: O for NGC 3379); Fried & Illingworth (1981: O for NGC 4472); Dressler (1980: ■). The King model is that of Fig. 2.

However, this is not the whole story for elliptical galaxies, as can be seen from Fig. 3. The situation is more complex. The data are plotted here in logarithmic coordinates to enable a direct comparison with the dispersion gradient in M87 (discussed below). Comparison can also easily be made in these coordinates between the dispersion profiles of the "normal" ellipticals, NGC 3379 and 4472, and the same King model dispersion profile as that used in Fig. 2. By so doing, it is clear that these two galaxies have dispersion profiles quite unlike NGC 4697 (for which $\sigma(r)$ is essentially flat at log $\sigma(r) \sim 2.2$) and are really rather consistent with the constant M/L isotropic models. Obviously there are differences in the trend of $\sigma(r)$ among ellipticals. In fact, if one looks at the available dispersion profiles for the 33 ellipticals, the sample splits very roughly at 50% flat, 50% decreasing $\sigma(r)$. Does this mean that some ellipticals show increasing M/L indicative of a massive halo component whereas some don't? Unfortunately, such a simple division may not be appropriate since, as mentioned above, interpretation of the dispersion profile in ellipticals is complicated by the fact that we are measuring the line-of-sight projection of essentially a single component of the velocity dispersion. There is no guarantee that all the components are equal. In fact, available data indicate that $\sigma_Z < \sigma_R \sim \sigma_\phi$, since the shape of ellipticals is determined primarily by anisotropy and not by rotation (Binney, 1978a). Thus it may well be that these $\sigma(r)$ profile differences are due to dynamical effects and not to real M/L changes.

The implications of the current $\sigma(r)$ data can best be understood as follows. Imagine the situation where one takes a constant M/L model like the King model in Fig. 3, but

31

allows anisotropy that systematically decreases σ_ϕ below σ_R with radius. Since the contribution of σ_ϕ to the <u>observed</u> dispersion σ_{los} increases with radius, one would observe a $\sigma_{los}(r)$ relation that fell <u>below</u> the constant M/L isotropic model with the deviation increasing with radius, the model and observed dispersion profiles being matched at the center. And vice versa if we increase σ_ϕ, compared to σ_R with radius. Clearly the introduction of such radial changes in σ_R or σ_ϕ influences the $\sigma_{los}(r)$ relation, and hence our interpretation of this relation.

Thus, given our data, where some $\sigma_{los}(r)$ = constant and some $\sigma_{los}(r)$ decrease, we have the following options for these two cases: (A) if $\sigma_{los}(r) \approx$ constant either (1) $\sigma_\phi(r)/\sigma_R(r) \sim$ constant and M/L increases, or (2) $\sigma_\phi(r)/\sigma_R(r)$ increases with r and M/L $\sim$ constant; (B) if $\sigma_{los}(r)$ decreases with r either (1) $\sigma_\phi(r)/\sigma_R(r) \sim$ constant and M/L $\sim$ constant, or (2) $\sigma_\phi(r)/\sigma_R(r)$ decreases with radius and M/L increases.

Of these, I find (A-2) unlikely in the context of typical collapse models of galaxy formation, where one would expect the velocity ellipsoid to point radially at large radii (e.g., see Binney [1980] and note the situation for RR Lyraes in the local solar neighborhood, where $\sigma_R^2 \sim 1.5\,\sigma_\phi^2 \sim 3\,\sigma_\theta^2$ [Woolley, 1978]). As always, though, one has a qualification. If ellipticals are formed through stellar mergers, then it may be that $\sigma_\phi > \sigma_R$ over some radius range in such systems. At the present time, however, the merger models do not have adequate statistics to study the velocity distribution function and so cannot indicate if this is likely.

Presently I would favor (A-1), i.e., $\sigma_{los}(r) \sim$ constant implies increasing M/L. In the case of (B), i.e., $\sigma_{los}(r)$ decreasing, no real case can be made for either (1) or (2) with current data, although Occam's razor may lead one to prefer (B-2), i.e., that M/L increases with radius in ellipticals in general, and that the degree of anisotropy varies among ellipticals. On the other hand, there are indications from some rotation curve data and from current binary galaxy data that not all disk galaxies have extensive, high M/L halos, so it is possible that some ellipticals may also lack such halos. One further caveat to all this is that it is only possible for anisotropy effects to dominate the $\sigma_{los}(r)$ relation over a limited radius range. Obviously at some point, if carried to extremes, all the dispersion comes to reside in one of the ϕ or R components, for example, and the mass distribution must again dominate the behavior of $\sigma_{los}(r)$.

A second point that can be made from the data in Fig. 3 is that there are "normal" ellipticals (NGC 3379, 4472) that show a trend in $\sigma(r)$ around r_c similar to that seen in M87, although there are indications from the latest data (symbol 0 in Fig. 3) for both of these normal galaxies that the slope may be less. The existence of such σ(r) behavior in the core of the "active" galaxy M87 has been attributed to the existence of a massive object in the nucleus (Sargent et al., 1978). If such σ(r) trends are also seen in some normal ellipticals, then we must either postulate the existence of like massive objects in the cores of these galaxies or must look for dynamical explanations, such as those suggested by Duncan and Wheeler (1980) and Binney (1978b).

The availability of velocity and velocity dispersion data for some 33 ellipticals enables us to define much more clearly the global rotational properties of these systems as described by the distribution in the $V/\sigma - \varepsilon$ plane shown in Fig. 4 (ellipticity $\varepsilon =$ 10 [1 - b/a], where b and a are the minor and major axes, respectively). As previously, ε is the maximum ellipticity inside the radius range covered by the rotation measurements. Binney (1980) has suggested that the flattening appropriate for this discussion is that at several core radii. Use of ellipticities from this region instead of maximum ellipticities moves the points toward lower ε, but the mean shift is small, being $\Delta\varepsilon \sim -0.05$. The velocity V_m is that of the flat portion of the rotation curve (e.g., see Fig. 2).

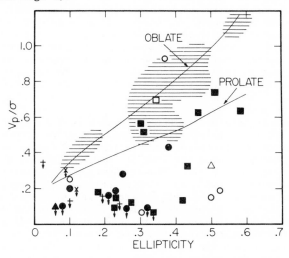

Fig. 4. The usual $V/\sigma - \varepsilon$ diagram for describing the global dynamical characteristics of stellar systems (details in the text). All individual points are elliptical data (except □ is M81 data) from several sources (see Fried & Illingworth, 1981). The models are those of Binney (1978a). The cross-hatched areas represent the distribution of bulge data discussed by Kormendy & Illingworth (1981) and Illingworth et al. (1981).

33

It is interesting to note that even though the rotation has little significance for the observed structure of the system, the form of the rotation curve is such as to invariably have a rapid rise to a maximum value (typically at several core radii) and to remain essentially constant out to the limits of the data — not unlike that expected for a disk in such systems. Also, the orientation of the angular momentum axis is such as to be aligned with the projected minor axis in almost all cases. Examples where there are considerable deviations between the photometric and kinematic axes also have morphological peculiarities (NGC 596: Williams, 1980; Schechter and Gunn, 1979; and NGC 4125: Bertola, this volume).

Finally, σ is taken to be a luminosity-weighted mean of the available data, unlike previous discussions where σ was the central value. This is more appropriate for comparison with the models discussed by Binney (1978a) that are plotted in Fig. 4, as well as eliminating any peculiarities associated with the cores of these galaxies.

The data can be compared directly with the plotted oblate model line, since the models have been projected. These oblate spheroid models have their equatorial plane in the line-of-sight and have velocity dispersions that are isotropic in their residual velocities. The flattening of these systems is accounted for by rotation. Random orientations of such models result, fortuitously, in points being distributed close to the above line (i.e., an E4 projected to look like E2 has the apparent rotation properties close to that of a true E2 seen in its equatorial plane). Clearly, only a very few ellipticals are consistent with such a model.

Again, the prolate models are defined as having flattening attributable to their rotation with isotropic dispersions. However, projection effects are nastier for this form. Imagine a prolate spheroid with both its long axis and its short rotation axis in the plane of the sky. Now if this object is rotated around its long axis, the shape will not change. However, in doing so, its rotation axis will go from being in the plane of the sky to being in the line-of-sight (these objects are tumbling end over end around a short axis, not rotating around the long axis); i.e., the rotation velocity seen will decrease from a maximum when we are looking through the equatorial plane to zero when we are looking along the rotation axis. Thus, if the galaxies were prolate, the V/σ observed in ellipticals seen at maximum axial ratio would range from zero to a maximum value near the oblate line. At smaller apparent ellipticities, the points would scatter over the V/σ-ε plane toward lower ellipticities as the orientation changed. Even an E0 could show "rotation" as a result of streaming motions in such a figure. Comparison with the data is

34

thus tricky. However, one can, as Binney (1978a) did, establish a
median line in this plane that would be consistent with the data
if ellipticals are generically rotating prolata. As can be seen
from Fig. 4, the available data are inconsistent with this
hypothesis. Unlike the rotating oblate case, however, one cannot
exclude on these grounds alone any single elliptical from having a
rotating prolate figure.

Clearly, as has been emphasized in the past, the solution lies
with anisotropy in the velocity dispersion tensor in the sense
that $\sigma_z < \sigma_R \sim \sigma_\phi$ if the generic form for ellipticals is an oblate
spheroid. In classical terms this would imply the existence of a
"third" integral beyond those of the energy and angular momentum.

While a simple anisotropic oblate form for ellipticals is a
distinct possibility, it is quite likely that, generically,
ellipticals are truly triaxial in form, as emphasized by Binney
(1976, 1978a) with support from recent models by Schwarzschild
(1979) and Wilkinson (1980) showing that such forms are, first,
possible and, second, stable on timescales approaching a Hubble
time. On quite general grounds one could well argue for such a
structure, given that anisotropy is clearly important in
establishing the structure of such systems. After all, since any
asymmetry in the protogalactic system, be it primordial or
resulting from tidal effects between protogalaxies (Binney & Silk,
1979), will grow during the protogalactic collapse era through the
mechanism discussed by Lin, Mestel & Shu (1965), there is no real
reason for imagining that initial or tidally induced anisotropies
were such as to lead to systems with two equal principal axes. It
is much more likely that all axes will differ, that the
asymmetries will grow by the Lin-Mestel-Shu mechanism, and that
any violent relaxation effects will not completely destroy the
resulting anisotropy (Aarseth & Binney, 1978), thus leaving a
triaxial figure.

Having seen that, generically, ellipticals are not systems in
which rotation plays any significant role in establishing their
present structure but are, in fact, anisotropy-dominated systems,
the question arises as to the structure of the elliptical-like
bulges of disk systems. The similarity of these systems, bulges
and ellipticals, in their stellar population and in their surface
brightness distributions would lead one to expect like dynamical
properties, would it not?

BULGES OF DISK SYSTEMS

In like manner to the studies of dispersion and rotation
profiles in ellipticals, a significant number of bulges have now
been investigated in an attempt to understand their dynamical
structure. In general, edge-on galaxies have been chosen so as to

minimize problems of disk contamination. The only exceptions have been for large, high surface brightness bulges where one can be sure of minimal disk contribution within the central kpc or so.

Data are now available for some 12 bulges, half of which are SOs, the remainder being Sa and Sb systems (Illingworth et al., 1981; Kormendy, 1981; Kormendy & Illingworth, 1981). An example of the data for this latter class, the Sb galaxy M81, can be seen in Fig. 5. The data shown reach to ~1.5 kpc along the major axis at which point the disk contribution is ~30% (Schweizer, 1976). Interior to ~0.5 kpc, we are clearly measuring the bulge rotation and dispersion, since the disk contribution is then $\lesssim$10%. Photometry (Brandt, Kalinowski & Roosen, 1972) shows this bulge to be ~E3.3. With $\langle\sigma\rangle$ ~ 150 km s^{-1} and V_m ~ 100 km s^{-1}, V/σ ~ 0.7, putting this object essentially on the rotating oblate spheroid line (Fig. 4: □).

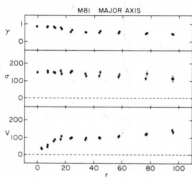

Fig. 5. As for Fig. 1 but for data from a major-axis spectrum of M81 (Verter, Illingworth & McElroy, 1981). The bulge dominates to beyond r ~ 60".

This is by no means an isolated case. Essentially all of the bulges studied show significant rotation — significant in the sense that $(V/\sigma)_{obs}/(V/\sigma)_{oblate}$ ~ 1. One of the most striking is NGC 3115 (Illingworth et al., 1981), a bulge-dominated, edge-on SO galaxy where the rotation, ~1 kpc up into the bulge, is still nearly 70% of the disk rotation of 270 km s^{-1}, and even at ~2.5 kpc above the disk the rotation is still 100 km s^{-1} (σ ~ 150 km s^{-1} throughout the bulge). This E5-E6 bulge (Strom et al., 1977) has V/σ ~ 1, nearly 2-4 times greater than for the few E5-E6 ellipticals in the sample shown in Fig. 4.

Similar data obtained for four Sa-Sb galaxies (NGC 4565, 4594, 5866, 7814) by Kormendy & Illingworth (1981) show large rotational velocities, also in the usual dimensionless sense. Of these, only the "Sombrero" NGC 4594, the galaxy with the bulge most like our elliptical sample in magnitude, has a V/σ that deviates significantly from the oblate line in Fig. 4. Even this object still falls among the top 20% of ellipticals in $(V/\sigma)_{obs}/(V/\sigma)_{oblate}$ terms.

GALAXY DYNAMICS: OBSERVATIONS

The bulge data currently available fall within the cross-
hatched areas in Fig. 4.   The appropriate place to site each
galaxy in the V/σ - ε plane is more difficult to establish for
bulges, at least in a fashion that is consistent with the
procedures used for the models and the ellipticals.   The reason
for this added difficulty is that the disk contaminates both the
light distribution and the velocity data near the equatorial plane
making it difficult to evaluate ε and V/σ for the bulge, ε and V/σ
both being overestimated if one is careless.   To minimize such
problems, we have used data at only small radii ($r \lesssim 1$ kpc) or at
high z distances, depending on the inclination of the object.

Clearly the bulge data in Fig. 4 are consistent with the
majority of the bulges, if not virtually all of them, being oblate
spheroids flattened by their rotation.   It is rather striking to
find that bulges (which at first glance are ellipticals with
disks) are rotation-dominated, considering that we have just
accustomed ourselves to thinking of ellipticals as anisotropy-
dominated systems.

In fact, the bulge data as plotted may <u>underestimate</u> the
importance of rotation.   The velocity used in V/σ is taken from
well above the disk in the edge-on galaxies (nearly half the
sample) and not from along the major axis as in the ellipticals
and as assumed by the models.   Clearly, the velocity obtained in
this way will be less than that measured along the major axis if
the disk were not present.   Elliptical data (Davies & Illingworth,
1981), models (e.g., Larson, 1975), and the bulge data itself show
clear gradients in rotation with z.   It is hard to estimate this
effect, but it may well lead us to underestimate V_m by 10-20%.

In addition, the observed ε of the bulge arises from two
effects: first, the flattening due to the dynamics of the bulge
(anisotropy, rotation), and the second, from the flattening of the
system's equipotentials by the inclusion of a disk.   This latter
effect is significant (Monet, Richstone & Schechter, 1981;
Freeman, 1980) but difficult to quantify.   Certainly removal of
the disk would lead to rounder systems, i.e., the cross-hatched
regions would move toward lower ε in Fig. 4, emphasizing even more
the importance of rotation in these systems compared to any
anisotropy in the stellar velocity distribution.

The bulges as a group differ from the elliptical sample so far
observed.   How much so can be seen in Fig. 6, where we have
plotted the distribution over $(V/\sigma)_{obs}/(V/\sigma)_{oblate}$ for the
ellipticals and the bulges in Fig. 4.   The difference is
striking.   So much so, in fact, that application of everybody's
favorite Russian test is not even necessary.

37

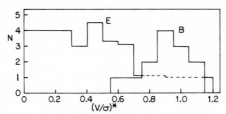

Fig. 6. The distribution over $(V/\sigma)* = (V/\sigma)_{obs}/(V/\sigma)_{oblate}$ for bulges B and ellipticals E. Objects that have $(V/\sigma)* \sim 1$ are consistent with being oblate spheroids flattened by their rotation.

There is a major question that arises, however, about the comparison in Fig. 7. The author is grateful to George Lake for emphasizing the importance of this point. We are to some extent playing the apples-oranges game in that the bulges are, in general, much less luminous than the ellipticals. This can clearly be seen in Fig. 7, where we show the distribution over magnitude for both samples. (The decomposition of L_{bulge} from L_{total} for the spirals and S0s was, of necessity, a very rough one — mostly eye estimates — "calibrated" by photometry from Burstein [1979] and Boroson [1980].) While rough, the mean is unlikely to be incorrect by 0.5 mag, certainly at most 1 mag; small compared with the differences apparent in Fig. 7.)

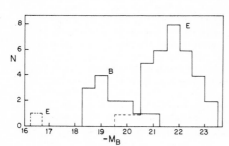

Fig. 7. The distribution over absolute magnitude (B_T from de Vaucouleurs, de Vaucouleurs & Corwin, 1976) for the elliptical E and disk system bulge B sample of Figs. 4 and 6. Simple estimates of bulge-to-disk ratios were used to derive the bulge magnitudes. The dotted E box at $M_B \simeq -16.5$ is M32.

The magnitude difference between the elliptical and bulge sample is large. M32, at $M_B \sim -16$ (or brighter if its rotation properties should really be attributed to the luminosity of its non-tidally limited progenitor [see Faber, 1973]) is the only elliptical fainter than -19, and it is a fence sitter having $(V/\sigma)_{obs}/(V/\sigma)_{oblate} \sim 0.5$. More velocity dispersion profiles and rotation curves are clearly needed to see if bulges and ellipticals of like luminosity ($M_B \gtrsim -20$) have like dynamics.

IMPLICATIONS FOR FORMATION

The outstanding results of the programs given here that should be addressed by any discussion of the formation of galaxies are:

GALAXY DYNAMICS: OBSERVATIONS

A. While the case for the existence of massive dark halos in early-type galaxies cannot be made in such unequivocal terms as that for later-type disk systems, indications are that such halos do exist in a significant fraction of elliptical and S0 systems.

B. Ellipticals as a class are anisotropy-dominated systems and are probably generically triaxial ellipsoids unless some process (mergers?) has rendered them axisymmetric. However, a significant fraction (~15%) do rotate fast enough for the rotation to have a significant effect on the structure of the galaxy.

C. Almost without exception, the bulges of spiral and S0 galaxies show considerable rotation and are consistent with their being genercially oblate spheroids flattened by their rotation.

This near dichotomy in the properties of ellipticals and the bulges of the disk systems is a very strong constraint on models of the galaxy formation process. One recent popular idea that seeks to explain this difference (e.g., Aarseth & Fall, 1980) considers ellipticals to be secondary objects, formed through the merger of typical disk systems. Doubts about the relevance of such a model existed initially because of the large rotations found in the remnants formed in model calculations of such mergers (White, 1979). This was, of course, at odds with the observational data on ellipticals. However, these doubts have, to some extent, been dispelled by recent numerical simulations of clustering in the early universe (Aarseth & Fall, 1980; Jones & Efstathiou, 1979) which indicate that most mergers occur between objects having very low orbital angular momentum.

Serious problems do exist with the merger hypothesis, however, as argued by Ostriker (1980) and others. Two problems of this hypothesis that I think are critical are the existence of line strength ($\equiv$ abundance) gradients in ellipticals over a very large range in radius (Faber, 1977), and the depth of the central potential in ellipticals compared with that in typical spirals.

It may be, however, that merging does have a role to play (it must, to some extent, because merging clearly occurs. But what do the remnants look like?). It is clear that dissipation has played a very important role in the formation of disk systems; obviously so in the case of the disk itself, but now also in the case of the bulge, as indicated by the data presented here. Abundance gradients and high central concentrations can be imagined to be readily established in such a dissipational collapse picture. However, such a collapse process generally results in an ordered velocity field (i.e., rotation), unlike that seen in ellipticals. Subsequent mergers of the stellar systems are excellent for mixing out such ordered motion, but have the disadvantages mentioned above. An attractive alternative approach discussed recently by Norman & Silk (1980) combines both of these features — dissipation and mergers — into a model whereby the bulk of the mergers occur while the systems are still essentially gaseous. This model is still at a very early stage of its development, but it shows promise in being able to generate a coherent picture of galaxy formation for all galaxy types.

While current theories of galaxy formation are still rather incomplete, this is an exciting period for those interested in formation processes because of the wealth of new data that is becoming available on the global properties of galaxies and the extensive modeling that is being carried out.

REFERENCES

Aarseth, S. J. & Binney, J. J. (1978). Mon. Not. R. astr. Soc., 185, p. 227.
Aarseth, S. J. & Fall, S. M. (1980). Astrophys. J., 236, p. 43.
Bertola, F. & Capaccioli, M. (1975). Astrophys. J., 200, p. 439.
Binney, J. J. (1976). Mon. Not. R. astr. Soc., 177, p. 19.
Binney, J. J. (1978a). Mon. Not. R. astr. Soc., 183, p. 501.
Binney, J. J. (1978b), private communication.
Binney, J. J. (1979). Phil. Trans. R. Soc. London, 296, p. 329.
Binney, J. J. (1980). Mon. Not. R. astr. Soc., 190, p. 421.
Binney, J. J. & Silk, J. (1979). Mon. Not. R. astr. Soc., 188, p. 273.
Boroson, T. A. (1980). Ph.D. thesis, University of Arizona.
Bosma, A. (1978). Ph.D. thesis, University of Groningen
Bosma, A. & Van der Kruit, P. C. (1979). Astron. Astrophys., 79, p. 281.
Brandt, J. C., Kalinowski, J. K., Roosen, R. G. (1972). Astrophys. J. Suppl., 24, p. 421.
Burstein, D. (1979). Astrophys. J., 234, p. 435.
Davies, R. L. (1980). Mon. Not. R. astr. Soc., in press.
Davies, R. L. & Illingworth, G. (1981). Astrophys. J., in preparation.
Davies, R. L., Illingworth, G. & McElroy, D. (1981) Astrophys. J., in preparation.
de Vaucouleurs, G., de Vaucouleurs, A. & Corwin, H. R. (1976). In Second Reference Catalogue of Bright Galaxies, Univ. of Texas Press, Austin.
Dressler, A. (1979). Astrophys. J., 231, p. 659.
Dressler, A. (1980). Astrophys. J., in press.
Duncan, M. J. & Wheeler, J. C. (1980). Astrophys. J. Lett., 237, p. L27.
Faber, S. M. (1973). Astrophys. J., 179, p. 423.
Faber, S. M. (1977). In The Evolution of Galaxies and Stellar Populations, Yale University Observtory, p. 157.
Freeman, K. C. (1980), in preparation.
Fried, J. & Illingworth, G. (1981). Astrophys. J., in preparation.
Gunn, J. E. (1979). Phil. Trans. R. Soc. London, 296, p. 313.
Illingworth, G. (1977). Astrophys. J. Lett., 218, p. L43.

Illingworth, G. & McElroy, D. (1981). _Astrophys. J._, in preparation.
Illingworth, G., Schechter, P. L. & Gunn, J. E. (1981). _Astrophys. J._, in preparation.
Jones, B. J. T. & Efstathiou, G. (1979). _Mon. Not. R. astr. Soc._, 189, p. 27.
King, I. R. (1966). _Astron. J._, 71, p. 64.
King, I. R. (1978). _Astrophys. J._, 222, p. 1.
Kormendy, J. (1981). _Astrophys. J._, in preparation.
Kormendy, J. & Illingworth, G. (1981). _Astrophys. J._, in preparation.
Larson, R. B. (1975). _Mon. Not. R. astr. Soc._, 173, p. 671.
Lin, C. C., Mestel, L. & Shu, F. H. (1965). _Astrophys. J._, 142, p. 1431.
Mathews, W. G. (1978). _Astrophys. J._, 219, p. 413.
Monet, D. G., Richstone, D. O. & Schechter, P. L. (1981). _Astrophys. J._, in press.
Mould, J. R., Illingworth, G. & Skillman, E. (1981). _Astrophys. J._, in preparation.
Norman, C. A. & Silk, J (1980). _Astrophys. J._, in press.
Ostriker, J. P. (1980). In _Comments on Astrophysics_, 8, p. 177.
Rubin, V. C., Ford, W. K., Jr. & Thonnard, N. (1980). _Astrophys. J._, 238, p. 471.
Rubin, V. C., Peterson, C. J. & Ford, W. K., Jr. (1980), _Astrophys. J._, in press.
Sargent, W. L. W., Schechter, P. L., Boksenberg, A. & Shortridge, K. (1977). _Astrophys. J._, 212, p. 326.
Sargent, W. L. W., Young, P. J., Boksenberg, A., Shortridge, K., Lynds, C. R. & Hartwick, F.D.A. (1978). _Astrophys. J._, 221, p. 731.
Schechter, P. L. (1980). _Astron. J._, 85, p. 801.
Schechter, P. L. & Gunn, J. E. (1979). _Astrophys. J._, 229, p. 472.
Schwarzschild, M. (1979). _Astrophys. J._, 232, p. 236.
Schweizer, F. (1976). _Astrophys. J. Suppl._ 31, p. 313.
Strom, K. M., Strom, S. E., Jensen, E. B. Moller, J., Thompson, L. A. & Thuan, T. X. (1977). _Astrophys. J._, 212, p. 335.
Terlevich, R., Davies, R. L., Faber, S. M. & Burstein, D (1980). _Mon. Not. R. astr. Soc._, in press.
Verter, F., Illingworth, G. & McElroy, D. (1981). _Astrophys. J._, in preparation.
White, S. D. M. (1979). _Mon. Not. R. astr. Soc._, 189, p. 831.
Wilkinson, A. (1980). Presented at NATO conference on "Normal Galaxies," Cambridge.
Williams, T. B. (1980). _Astrophys. J._, in press.
Woolley, R. (1978). _Mon. Not. R. astr. Soc._, 184, p. 311.

SHAPES OF UNPERTURBED GALAXIES

Martin Schwarzschild

Princeton University Observatory

1. VARIETY OF EQUILIBRIUM CONFIGURATIONS

I would like to restrict myself here to unperturbed galaxies, i.e. galaxies without a nearby important companion and without spiral structure of large scale and amplitude. Astronomers like H.C. Arp and A. Toomre might counter that in this manner I limit myself to a minority of galaxies, generally of unimpressive looks. I would fully agree and even emphasize that the study of clean perturbations, spiral or tidal or both, is sure to enhance greatly our understanding of galaxies —just as the study of Stark and Zeeman effects did for atoms and of pulsations and apsidal motions for stars. Attention to unperturbed galaxies might, however, shed light on a decisive question: Is the shape of a galaxy, in the absence of active perturbations, dominated by present equilibrium conditions or by the character of the origin? In either alternative we can learn much —but only if we can guess which alternative is the more relevant.

The approaches to this question may be roughly divided into two classes. In the first class the emphasis is on the processes governing the formation of a galaxy, and the evolution from origin to equilibrium is generally followed by large-scale computations. This approach has not only given us the first glimpses regarding the formation of galaxies but has already provided us with indications of a wider variety of equilibrium shapes than we might previously have contemplated (Aarseth & Binney, 1978; Miller & Smith, 1979). In the second class attention is entirely concentrated on equilibrium configurations with the explicit aim of exploring the variety of possible shapes. Here studies have mainly employed analytical methods (Lynden-Bell, 1962b; Freeman, 1965; Vandervoort, 1980) but recently also numerical procedures (Schwarzschild, 1979; Merritt, 1980; Richstone, 1980).

The main intent of this presentation is to suggest that the variety of equilibrium shapes for big stellar systems is likely to be large and that accordingly we can learn from detailed observations of unperturbed galaxies significantly more about their origin than we thought in the past.

2. LEVELS OF SYMMETRY

For the purpose of orientation let us consider three levels of symmetry: spherical, axial and triaxial. In this sequence each level permits a larger variety of shapes than the preceding one.

Observations suggest the existence of stellar systems of all three of these symmetry levels. The majority of globular clusters are clearly well fitted by spherical symmetry; So also are some of the EO galaxies though others may owe their circular appearance on the sky to accidents of projection. Axial symmetry has been the normal presumption for the basic shape of major galaxies. It continues to be a natural guess for systems dominated by rotation, such as galactic disks. Observational indications for triaxial shapes are less direct. For the bulge of the Andromeda Nebula Lindblad (1956) has early pointed out that its twist, i.e. the deviation in position angle of its apparent major axis from that of the disk, can be understood if a bar-like triaxial form is assumed. Similar twists, with similar interpretations, have more recently been found for many elliptical galaxies (King, 1978; Strom & Strom, 1978; Williams & Schwarzschild, 1979a,b). In parallel, a dynamical argument for triaxiality is based on the observation (Illingworth, 1977) that rotation plays only a minor role for most elliptical galaxies. If this observation forces us to introduce a strongly anisotropic velocity distribution (Binney, 1980a) it would seem as rational to assume a triaxial distribution as an axially symmetric one.

What then are the basic differences between the three symmetry levels for the dynamics of stellar systems? Let us pursue this question by the following rough approach. We choose an arbitrary density distribution of the symmetry under consideration and ask whether there will generally exist a phase-space distribution function, f, which is self-consistent with the density and the corresponding potential. Our answer -rough indeed- will depend solely on the number of integrals of motion which exist for the potential and hence can be used as independent arguments of f.

At the first symmetry level, a spherical density distribution depends only on one coordinate while the corresponding spherical potential has four classical integrals. Hence an arbitrary spherical density can be expected to have a multitude of self-consistent f's -a disturbing degree of underdeterminacy. For the main example of this symmetry, the globular clusters, the character of the problem is, however, entirely altered by the fact that for them -in contrast to normal galaxies- stellar encounters are of decisive importance. The study of spherical systems with encounters is extensive and quite distinct from our main topic, the dynamics of systems with a variety of shapes and without encounters.

SHAPES OF UNPERTURBED GALAXIES

At the second symmetry level, an axial density distribution depends on two coordinates, and the corresponding axial potential has two classical integrals. This coincidence encourages us to guess that an arbitrary axial density distribution may in general have a self-consistent f. Guessing, however, is not necessary for this conclusion because of Lynden-Bell's (1962a) study of this problem.

This is not the end of the story of axial symmetry; a serious discrepancy did plague this field. An f function depending on the two classical integrals for axial symmetry will always produce equal velocity dispersions in the radial and axial directions -in striking contrast to observations for our Galaxy in the solar neighborhood. This dilemma was approached in very different ways by two giants of our discipline, J. Oort and S. Chandrasekhar, a fascinating occurrence for future science historians. The dilemma was resolved by the development of a third effective integral for the potential of the Galaxy by Contopoulos (1959). By now we know that generally potentials relevant to galaxies with axial symmetry have such a third integral. Thus also this symmetry level presents, in our terms, an underdetermined problem. For an arbitrary axial density distribution the equilibrium conditions are not enough to fix its dynamical state. An axial galaxy has the capability of retaining details of conditions at its birth -a clear help to studies of galaxy formation.

At the third symmetry level, a triaxial density distribution depends explicitly on all three coordinates. But the corresponding potential has only one classical integral, the energy or the Jacobi constant for the case of a rotating figure. Our rough argument then leads to the statement that self-consistent tri-axial equilibrium configurations can exist only if the relevant potentials have two additional nonclassical effective integrals. Such a sweeping estimate obviously has to be fenced in with precautions. First, there are the special potentials listed by Lynden-Bell (1962b) which have explicit additional integrals that one might want to group with the classical ones. Indeed for one of these potentials, the one corresponding to uniform density within the figure, Freeman (1965) has derived an exact triaxial solution. Second, for another set of potentials Vandervoort (1980) has found special exact solutions in which f depends solely on the one classical integral. But for both these cases fairly special conditions have to be fulfilled. I would therefore suspect them of being examples of special solutions which altogether comprise only a limited portion of all three-dimensional equilibrium configurations relevant for galaxies. I would like then to proceed under the assumption that for a triaxial equilibrium configuration the existence of two additional effective integrals is -in general- necessary.

3. NON-CLASSICAL EFFECTIVE INTEGRALS

I would like to use a definition for an effective integral which is directly related to its expected role in an equilibrium model for a galaxy. We may thus call an effective integral any isolating function of the six phase-space coordinates which is substantially constant on an orbit for the relevant time. The required level of constancy we may describe, for example, by the condition that three effective integrals together should fix the velocity vector at a point in ordinary space to within, say, one-percent of the escape velocity from the center. For the time span relevant for a galaxy we naturally choose the Hubble time.

This restriction of the time span to be considered brings out a major difference between the dynamics of a galaxy and that of the solar system (and equivalent situations in plasma experiments and accelerators). In a galaxy the dynamical time is of the order of 10^7 or 10^8 yrs. (excluding galactic nuclei with dimensions of a few parsecs). With the Hubble time of about 10^{10} yrs. the decisive ratio of lifetime to dynamical time is at most 10^3. For the solar system the same ratio is roughly 5 x 10^9 yrs/5 yrs. = 10^9. Hence for the solar system deep and intricate processes have ample time potentially to develop significant consequences while the same danger seems minimal for a galaxy. Furthermore, the great difference in the decisive ratio has a parallel practical consequence. Modern computers still seem insufficient to tackle dynamical problems regarding developments in the solar system over its entire life span. In contrast it is a minor task for present computers to follow the orbit of a star in a galactic potential for a Hubble time. Accordingly, it is numerical experiments regarding individual orbits -not N-body calculations- on which most of the following discussion is based.

Among the effective integrals as here defined the classical integrals can be expressed in closed form and the formal integrals by truncated series developments (Contopoulos, 1959). It is those nonclassical effective integrals which are not easily represented by a short development that make the most trouble. But it is exactly these integrals which appear to play a major role for tri-axial figures with strong central density concentration simulating those of galaxies. Luckily for the specific approach here under discussion, a representation is not required for these troublesome integrals. Only assurance of their existence is needed for any specific case.

For the demonstration of the existence of an additional integral in a two-dimensional system the surface of section has proven a most effective tool (Henon & Heiles, 1964). This technique has been particularly fruitful for galaxy models with axial symmetry for which the equations of motion can be transformed into the co-moving meridional plane (Ollongren, 1965). For three-dimensional systems equivalent though much more cumbersome techniques can be used, such as the following. Select an arbitrary point in the figure and record the velocity vector every time the orbit to be tested for additional effective integrals passes near the selected point. Because of the energy integral these vectors must form a sphere in velocity space. If an orbit has no additional effective integral the velocities will fill an area on the sphere. If the orbit has one additional effective integral the velocities will fall along a curve on the sphere, and with two additional effective integrals all such velocity vectors will be identical, i.e. fall on a point on the sphere —except for a finite multiplicity, normally fourfold or eightfold.

Investigations using these and similar techniques have shown for several axial and triaxial models which fit observed density profiles that many orbits have three effective integrals. Indeed I feel that it is now a reasonable guess that most axial and tri-axial potentials relevant for galaxies have a majority of orbits with three effective integrals. However uncertain this guess may still appear, it sure is now safer than the opposite guess that effective integrals other than the classical ones are a rarity.

Are triaxial shapes the end of the line? Could still more com-plicated shapes such as mildly twisted bars also be equilibrium configurations? Those are questions for the future. Let us for the rest of this presentation concentrate on triaxial models.

4. NUMERICAL CONSTRUCTION OF EQUILIBRIUM MODELS

I would like to sketch briefly here a numerical procedure for the construction of self-consistent equilibrium models. The pur-pose of this sketch is to bring out the role played by regular orbits, i.e. orbits with three effective integrals. The technical details of the procedure (Schwarzschild, 1979) are here not rele-vant.

To start the procedure choose a density distribution. For the time being restrict your choice to one with triaxial symmetry. You may fit your choice to an actual galaxy —one that is not actively perturbed from equilibrium. Next, derive the corresponding po-tential from the Poisson equation. Third, compute an ample, representative sample of individual orbits in this potential. Run each orbit for, say, a hundred oscillations and record the density distribution which it produces in three-dimensional space. Finally, try to reconstruct your choice for the total density dis-tribution of the model by adding up all the individual orbital density distributions, after occupying each orbit with an appro-priate number of stars. Remember that these occupation numbers must not be negative, i.e. use linear programming. If you succeed with this last step you have constructed a numerical self-consis-tent equilibrium model —with all the uncertainties and inaccuracies inherent in such a construction.

Obviously, the chance of success in the final step of this procedure is bigger the greater the variety one finds among the individual orbital density distributions. Equally obviously, the variety in the orbital density distributions will depend decisively on the number of effective integrals provided by the chosen potential. Less than three effective integrals will generally not suffice for a truly three-dimensional density distribution such as a triaxial one.

While the sketched procedure exploits the existence of additional effective integrals it does not require their isolation or description in any form.

Experience with this procedure is still minimal (Richstone, 1980; Schwarzschild, 1979). Nevertheless, it encourages the belief in the existence of equilibrium configurations with a triaxial shape and a central density concentration as high as that observed in galaxies.

Existence of the sought-for configuration is the main point of this inquiry, whereas uniqueness of the solution for the problem as here formulated seems of less immediate astronomical relevance. We want to know whether there exists an equilibrium configuration for a prescribed density model of complicated shape; it would not disturb our main pursuit if we found more than one such configuration for the same density model. On the other hand, the question of stability is highly relevant -and largely unanswered- for numerical equilibrium models. The stability of one model derived by the procedure just sketched has recently been tested (Miller & Smith, 1980) by a large N-body calculation. It was found that this triaxial model is stable against catastrophic changes, but the available numerical accuracy was insufficient to test against slower instabilities. A quite different scheme to test for slow instabilities has been suggested (Binney, 1980b) but has not yet been applied. Clearly, the problem of stability looms at least as large for numerical models as for analytical models.

5. STOCHASTIC ORBITS

It is time in this discussion to turn to the unpleasant topic of orbits which have no effective integrals other than the energy integral, i.e. stochastic orbits. Until now I have emphasized the experience that in many potentials relevant to galaxies the majority of orbits have three effective integrals, i.e. are regular. But the existence of a minority of orbits that are stochastic is an equally frequent experience. In a surface of section the stochastic orbits are commonly found to occupy a fraction of the area inside the energy limit high enough to worry about.

In a surface of section for a two-dimensional system a stochastic orbit can not penetrate into an area occupied by the curves representing regular orbits. The same is not exactly true for a three-dimensional system. However, the diffusion of stochastic orbits into an area densely occupied by regular orbits appears to be generally so slow in such a system that it should be ignorable for galaxies because of their low decisive ratio of lifetime to dynamical time (at most 10^3). Thus for both two- and three-dimensional systems we may consider the areas occupied by stochastic orbits as reasonably distinct from those occupied by regular orbits –notwithstanding the circumstance that the boundaries are prone to complicated fine structure.

One might reasonably expect that a stochastic orbit would fill all phase space available to it, constrained only by its energy integral and the existence of regular orbits. If this is true, an entire stochastic area contains only one orbit and would accordingly provide only one orbital density distribution for the procedure sketched in the preceding section. Such a reduction in the available orbital density distributions by truly stochastic orbits does imply limitations on the model density distributions for which equilibrium configurations exist.

The expectation of truly stochastic behavior of non-regular orbits has been put into doubt by a recent numerical experiment (Goodman & Schwarzschild, 1980). This experiment is limited to a specific triaxial potential with a Hubble profile for the density and with a figure at rest in an inertial frame. It is furthermore limited to just four orbits, all in the same stochastic area at one energy. The orbits were computed for a Hubble time. Each of these orbits showed a stochastic behavior: in a surface of section –as described in §3– its velocity vectors scatter over an area, rather than falling on a point as would be the case for a regular orbit. But its behavior is not truly stochastic: the area over which the velocity vectors of one orbit scatter is substantially smaller than the total stochastic area to which the orbit belongs. But the most striking deviation of these orbits from the behavior expected for truly stochastic orbits is that they produced orbital density distributions strongly and systematically differing from one another.

In spite of the narrow limits of this numerical experiment its results encourage the estimate that the appearance of stochastic orbits may not imply as much of a limitation on the shapes of equilibrium models as one might have thought.

6. MAJOR FAMILIES OF REGULAR ORBITS

The structure of an equilibrium model for a galaxy can be approached either by the explicit use of the collisionless Boltzmann equation or by the consideration of an assembly of stars occupying a variety of orbits in a given potential. In the work here reported the latter approach has been used and we shall exploit it further in this last section.

The variety of orbits in a three-dimensional potential is large, given the existence of three effective integrals. Luckily this confusing variety can be sorted out in many cases into distinct families of orbits, each surrounding a parent orbit which is closed and hence periodic. Still the number of closed orbits is large in a general potential and so is the number of corresponding orbit families. However, experience with some axial and triaxial potentials suggests that commonly all but a modest fraction of phase space will be occupied by orbits belonging to a few major families surrounding the most basic closed orbits. If one then permits oneself the neglect of the multitude of minor orbit families which surround all the more complicated closed orbits, then the variety of the remaining orbits, i.e. those belonging to the major families, may be both manageable and sufficient for our purposes.

Let us exemplify this approach by a brief discussion of three classes of systems: cold axial systems, hot axial ones and hot triaxial ones. Cold systems are defined as configurations in which the mean stream motions provide the acceleration caused by the gravitational potential while the velocity dispersions are small compared to the mean motions. In contrast, in a hot system the velocity dispersions dominate and any mean motions are secondary.

In a potential with axial symmetry two types of basic closed orbits exist, radial orbits and circular ones in the equatorial plane. In a cold system the families around radial orbits have to be unoccupied -which incidentally relieves us from having to worry about their possible stochasticity. A radial orbit will provide at any location an equal number of stars going outwards and going inwards. Thus it produces a velocity dispersion of the same order as the circular velocity, in contradiction with the definition of a cold system. Thus for a cold axial model we need consider only the major family of orbits which surrounds the circular one.

This class of models has two enormous advantages. First, it fits the majority of stellar orbits in our Galaxy near the sun, as well as those in all thin galactic disks. Second, it is easily accessible to analytical treatment, an advantage exploited at the very start of galaxy dynamics by B. Lindblad and J. Oort. This latter advantage arises from the fact that a near-circular orbit occupies only a portion of the meridional plane small enough for the potential to be well represented in analytical form. The same advantage was also exploited for the development of the "third" integral (Contopoulos, 1959), the first non-classical effective integral in a galactic potential. Recently the analytical advantage of a cold disk has even been used to investigate bar-like distortions from axial symmetry (Contopoulos, 1979; Lynden-Bell, 1979), with results relevant for cold bars -in distinction to hot triaxial configurations.

For the second class of systems, hot models with axial symmetry, a substantial literature exists concerning two special cases: first, potentials for which the third integral is available in analytical form and, second, general potentials for which solutions are sought with the phase-space distribution function depending only on the two classical integrals which always exist for this symmetry level (e.g. Hunter, 1977). The exploitation of a third effective integral in the general case of this class is complicated since the dominant orbits cover large regions in the potential so that analytical developments seem rather less practicable than for the cold class. However, a hot axial model which strongly, though implicitly, exploits the third effective integral has recently been derived by the numerical procedure sketched in §4 (Richstone, 1980). Such models may well be relevant to elliptical galaxies, which according to modern observations (Illingworth 1977) seem generally to be hot systems.

Finally, for the class of hot triaxial systems analytical approaches appear to be exceedingly difficult, except for some special potentials (Lynden-Bell, 1962b; Freeman, 1965; Vandervoort, 1980) and for the near-harmonic cores of any non-singular potentials (De Zeeuw, 1979). Here then numerical methods exploiting the existence of major orbit families may have their place. Owing to my inexperience with rotating figures the remaining discussion is restricted to figures fixed in inertial space.

For triaxial figures six simple closed orbits may be considered as parent orbits for orbit families, one along each of the principal axes and one circling each of these axes in a symmetry plane perpendicular to it. The families surrounding the three axial orbits may be combined into one family of three-dimensional box orbits, i.e. orbits with stationary points (distinct from two-dimensional box orbits in the meridional plane of an axial system, orbits which correspond to three-dimensional tube orbits). The family of box orbits will generally contain some stochastic orbits with stationary points near the minor axis (Goodman & Schwarzschild 1980). But most importantly it contains an ample variety of regular orbits with three effective integrals, all surrounding the major axis, i.e. just the type of orbits needed for a triaxial equilibrium figure.

Of the three circling parent orbits the one around the intermediate axis is unstable and does not produce a family of tube orbits (Heiligman & Schwarzschild, 1979). But the circling closed orbits around the major and minor axes are generally stable and each produce a major family of tube orbits. Among the tube orbits around the major axis there are many whose largest extent lies in the direction of this axis, just right to join the appropriate box orbits in the support of the triaxial figure.

This qualitative review of the major orbit families in a tri-axial potential, bolstered by a numerical triaxial model with a density distribution taken from observations (Schwarzschild, 1979; Merritt, 1980), encourages the estimate that hot triaxial equili-brium configurations do exist and that they may be relevant to many elliptical galaxies as suggested by their observed apparent twists. The more far-reaching question, whether there exist equi-librium figures of still more complicated shapes, appears to re-main entirely open.

This report is based less on reading the relevant literature than on many discussions with colleagues, most of all with Dr. James Binney. This work was supported by NSF Grant AST78-23796.

REFERENCES

Aarseth, S.J. & Binney, J. (1978). On the relaxation of galaxies and clusters from aspherial initial conditions. Mon. Not. Roy. Ast. Soc., 185, pp. 227-243.

Binney, J. (1980a). The dynamics, shapes and origins of elliptical galaxies. Phil. Trans. R. Soc. London, A296, 329-338.

Binney, J. (1980b). Private communication.

Contopoulos, G. (1959). A third integral of motion in a galaxy. Zs. f. Ap., 49, pp. 273-291.

Contopoulos, G. (1979). How far do bars extend? ESO Sci. Prepr., No. 53, pp. 1-48.

De Zeeuw, T. (1979). Private communication.

Freeman, K.C. (1965). Structure and evolution of barred spiral galaxies, II. Mon. Not. Roy. Ast. Soc., 134, pp. 1-23.

Goodman, J. & Schwarzschild, M. (1980). Semistochastic orbits in a triaxial potential. Ap.J. Suppl., submitted.

Heiligman, G. & Schwarzschild, M. (1979). On the nonexistence of three-dimensional tube orbits around the intermediate axis in a triaxial galaxy model. Ap.J., 233, pp. 872-876.

Henon, M. & Heiles, C. (1964). The applicability of the third integral of motion: Some numerical experiments. A.J., 69, pp. 73-79.

Hunter, C. (1977). Relation between the dynamics and the flat-tening of elliptical galaxies. A.J., 82, pp. 271-282.

Illingworth, G. (1977). Rotation (?) in 13 elliptical galaxies. Ap.J. Letters, 218, pp. L43–L47.

King, I.R. (1978). Surface photometry of elliptical galaxies. Ap.J., 222, pp. 1–13.

Lindblad, B. (1956). On a barred spiral structure in the Andromeda Nebula. Stockholm Obs. Ann., Vol. 1, No. 2, pp. 1–12.

Lynden-Bell, D. (1962a). Stellar dynamics: Exact solution of the self-gravitation equation. Mon. Not. Roy. Ast. Soc., 123, pp. 447–458.

Lynden-Bell, D. (1962b). Stellar dynamics: Potentials with isolating integrals. Mon. Not. Roy. Ast. Soc., 124, pp. 95–123.

Lynden-Bell, D. (1979). On a mechanism that structures galaxies. Mon. Not. Roy. Ast. Soc., 187, pp. 101–107.

Merritt, D. (1980). A numerical model for a triaxial stellar system in dynamical equilibrium. II. Some dynamical features of the model. Ap.J. Suppl., 43 (in press).

Miller, R.H. & Smith, B.F. (1979). Six collapses. Ap.J., 227, pp. 407–414.

Miller, R.H. & Smith, B.F. (1980). Private communication.

Ollongren, A. (1965). Theory of stellar orbits in the Galaxy. Ann. Rev. Astr. and Ap., 3, pp. 113–134.

Richstone, D. (1980). Scale-free, axisymmetric galaxy models with little angular momentum. Ap.J., 238 (in press).

Schwarzschild, M. (1979). A numerical model for a triaxial stellar system in dynamical equilibrium. Ap.J., 232, pp. 236–247.

Strom, K.M. & Strom, S.E. (1978). Surface brightness and color distributions of elliptical and S0 galaxies. I. The Coma cluster elliptical galaxies. A.J., 83, pp. 73–133.

Vandervoort, P.O. (1980). The equilibrium of a galactic bar. Ap.J., 240 (in press).

Williams, T.B. & Schwarzschild, M. (1979a). A photometric determination of twists in early-type galaxies. Ap.J., 227, pp. 56–63.

Williams, T.B. & Schwarzschild, M. (1979b). A photometric determination of twists in early-type galaxies, II. Ap.J. Suppl., 41, pp. 209–213.

THE MACROSCOPIC DYNAMICS OF ELLIPTICAL GALAXIES

James Binney

Princeton University Observatory

Professor Schwarzschild has described several studies of the microstructure of elliptical galaxies – how individual stars move through a system, and thus how the overall density structure is built up out of a myriad of individual orbital densities. It is impossible not to be impressed by the power and flexibility of this approach. Yet one must also stress the labour involved in constructing model galaxies on this plan. Fortunately one may obtain some understanding of elliptical galaxies from macroscopic considerations, that is, without asking explicitly which stars move where, but rather focusing on the collisionless Boltzmann equation and its associated moment equations.

EQUATIONS OF MACROSCOPIC DYNAMICS

Elliptical galaxies are sufficiently large and hot that the (two-body relaxation) time-scale on which energy is exchanged between stars is longer than the Hubble time; anything sufficiently small and compact that this condition fails is commonly dismissed as mere star-cluster. Therefore the density $f(\underline{x},\underline{v})$ of stellar mass in six-dimensional phase space $(\underline{x},\underline{v})$ obeys the collisionless Boltzmann equation

$$\frac{\partial f}{\partial t} + v_i \frac{\partial f}{\partial x_i} - \frac{\partial \Phi}{\partial x_i} \frac{\partial f}{\partial v_i} = 0 \qquad (1)$$

If one multiplies this equation by the kth component of velocity, v_k, and integrates over all velocities, one obtains the basic equation of 'stellar hydrodynamics':

$$\frac{\partial \rho \overline{v}_k}{\partial t} = - \frac{\partial (\rho \overline{v_k v_i})}{\partial x_i} - \rho \frac{\partial \Phi}{\partial x_k} \quad , \qquad (2)$$

where

$$\rho(\underline{x}) \equiv \int f(\underline{x},\underline{v}) \, d^3 v \qquad (3a)$$

55

J. BINNEY

$$\overline{v}_k \equiv \frac{1}{\rho} \int f \, v_k \, d^3v \tag{3b}$$

$$\overline{v_k v_i} \equiv \frac{1}{\rho} \int f \, v_k v_i \, d^3v \ , \tag{3c}$$

and the second term on the right hand side of equation (2) has been reformed by an integration by parts in velocity space. Equation (2) relates the rate of change of stellar momentum $\rho\underline{v}$ to the gradient of the kinetic energy density $\frac{1}{2}\rho\overline{v_k v_i}$, which it is often useful to decompose into components associated with stellar 'pressure' and with streaming motions. The equation of continuity of the stellar 'fluid' follows immediately from (1) by integration over all velocities:

$$\frac{\partial \rho}{\partial t} + \frac{\partial}{\partial x_i} (\rho \overline{v}_i) = 0 \tag{4}$$

I shall return to these hydrodynamic equations, but wish now to press on to a higher order moment of equation (1); that which enshrines the tensor virial theorem. Multiply (2) by x_j and integrate over some volume of space τ;

$$\frac{d}{dt} \int_\tau x_j \rho \overline{v}_k d^3x = -\int_{\partial\tau} x_j \rho \overline{v_k v_i} \, dS_i + \int \delta_{ji} \rho \overline{v_k v_i} d^3x - \int \rho x_j \frac{\partial \Phi}{\partial x_k} d^3x$$

$$= \tilde{S}_{jk} + 2T_{jk} + \Pi_{jk} + W_{jk} \tag{5}$$

where

$$\tilde{S}_{jk} \equiv -\int_{\partial\tau} \rho x_j \overline{v_k v_i} dS_i \tag{6a}$$

$$T_{jk} \equiv \frac{1}{2}\int \rho \overline{v}_j \overline{v}_k \, d^3x \tag{6b}$$

$$\Pi_{jk} \equiv \int \rho (v_j - \overline{v}_j)(v_k - \overline{v}_k) d^3x \tag{6c}$$

$$W_{jk} \equiv -\int \rho x_j \frac{\partial \Phi}{\partial x_k} \, d^3x \tag{6d}$$

The left hand side of equation (5) can be brought to a more intuitive form by defining a kind of moment of inertia tensor I_{jk} (not quite the usual one) by

$$I_{jk} \equiv \int \rho x_j x_k d^3x \tag{7}$$

and observing that

$$\frac{1}{2} \frac{dI_{jk}}{dt} = \int \frac{\partial \rho}{\partial t} x_j x_k \, d^3x = -\frac{1}{2}\int \frac{\partial}{\partial x_i} (\rho \overline{v}_i) x_j x_k \, d^3x$$

$$= \frac{1}{2}\int \rho (\overline{v}_j x_k + \overline{v}_k x_j) \, d^3x \ , \tag{8}$$

where I have used equation (4) and integrated once by parts. Evidently $\frac{1}{2}d^2I_{jk}/dt^2$ equals the symmetric part of the left hand side of equation (5). T and Π are clearly symmetric tensors, and it is not difficult to show (Chandrasekhar 1968) that for a self-gravitating system W is also symmetric. Therefore if we define $S_{jk} = \frac{1}{2}(\tilde{S}_{jk} + \tilde{S}_{kj})$ to be the symmetric part of $\tilde{S}$, we have that

$$\frac{1}{2}\frac{d^2I_{jk}}{dt^2} = S_{jk} + 2T_{jk} + \Pi_{jk} + W_{jk} \ . \tag{9}$$

The familiar scalar virial theorem is the trace of this equation.

APPLICATION OF THE TENSOR VIRIAL THEOREM

The tensor W depends only on the geometry of the galaxy's density distribution. If we refer everything to the principal axes of the galaxy, only the three diagonal components of W are non-zero; all components of W are negative, and if we let the longest and shortest axes of the galaxy define the x- and z-directions respectively, then $|W_{xx}| \geq |W_{yy}| \geq |W_{zz}|$. If the figure of the galaxy is fixed in space, $d^2I/dt^2 = 0$, and the physical content of equation (9) is that there must be more kinetic energy in the x-direction than in the z-direction. This extra motion may take the form of organized streaming motion (large T_{xx}) or an aniso-tropic pressure ($\Pi_{xx} > \Pi_{zz}$); it may occur near the centre or at great radii. But extra x-motion there must be if equation (9) is to be satisfied.

If the figure of the galaxy rotates steadily with angular velocity ω about the z-axis, the non-zero components of (9) become

$$2T_{xx} + \Pi_{xx} + \omega^2\delta I + W_{xx} = 0 \tag{10a}$$

$$2T_{yy} + \Pi_{yy} - \omega^2\delta I + W_{yy} = 0 \tag{10b}$$

$$2T_{zz} + \Pi_{zz} + W_{zz} = 0 \tag{10c}$$

where $\delta I \equiv I_{xx} - I_{yy} \geq 0$. According to these equations, rotation of the figure of the galaxy adds an effective kinetic energy $\omega^2\delta I$ to the longer axis of the equatorial plane, and removes an equal kinetic energy from the shorter equatorial axis. This effect enables systems whose pressures are isotropic ($\Pi_{xx} = \Pi_{zz}$) and dominated by rotation ($T_{xx} \simeq T_{yy} \gg T_{zz}$) to be tri-axial (Vandervoort 1980).

Define four new quantities $\bar{v}$, $\bar{\sigma}$, δ and q by $M\bar{v}^2 = 2(T_{xx}+T_{yy})$, $M\bar{\sigma}^2 = \Pi_{xx}$, $\delta = (\Pi_{xx} - \Pi_{zz})/\Pi_{xx}$ and $q = \Pi_{yy}/\Pi_{xx}$, where M is the total mass of the system. Physically, $\bar{v}$ may be interpreted as the r.m.s. rotational velocity in the (x,y) plane, $\bar{\sigma}$ is the r.m.s. random velocity along the x axis, δ is the mean fractional excess of pressure in the x- as against the z-direction and q is the ratio of the pressures in the y- and x-directions. Introduce these quantities into (10) to replace T and Π, set $T_{zz} = 0$ (which amounts to discounting the possibility of meridonal circulation) and divide the sum of (10a) and (10b) by (10c) to obtain

$$(\overline{v}/\overline{\sigma})^2 = \left[(W_{xx} + W_{yy})/W_{zz}\right] (1 - \delta) - (1 + q) . \tag{11}$$

If one restricts oneself to the simple case of galaxies whose iso-density surfaces are all similar to one another, one may employ a beautiful result from potential theory which states that $(W_{xx}+W_{yy})/W_{zz}$ depends only on the axial ratios (Roberts 1962). That is, the radial density profile factors out of the final answer. Therefore, $\overline{v}/\overline{\sigma}$ is under these circumstances completely determined by the axial ratios and the anisotropy parameters δ and q.

If the galaxy is biaxial and rotates about its axis of symmetry, one has that q = 1 by symmetry, so $\overline{v}/\overline{\sigma}$ depends only on δ and the ellipticity ε. Figure 1a shows this relationship for oblate galaxies after the appropriate projection of $\overline{v}$ along a line of sight in the equatorial plane. Figure 1b shows an analogous result for prolate galaxies that rotate about one of their minor axes. If all ellip-tical galaxies were prolate spheroidal, had isotropic velocity distributions and rotated about one of their minor axes, half of their representative points in a $\overline{v}_p/\sigma$ versus ε diagram would lie below the dashed line shown in Figure 1b, and half above this line The position of the representative point of any particular galaxy in this diagram depends on its orientation to the line of sight. When viewed down the z-axis, so that it appears highly elongated and yet without rotation, its representative point in Figure 1b will lie along the horizontal axis of the diagram; when viewed along the x-axis the galaxy will appear round and rapidly rotating so that its point in Figure 1b will lie on the vertical axis, and when the system is viewed down the y-axis, so that it appears to be maximally elongated and maximally rotating, its representative point will lie along the full line in Figure 1b (Binney 1978).

These results are exact and rather general, but they do not lend themselves to direct confrontation with observation. Indeed, if one number characterizes the rotation of a galaxy, it is v_m/σ_c, the ratio of the peak rotation velocity to the central line of sight velocity dispersion. Figures 1a and 1b show predictions for the ratio of a mean rotation velocity to a mean velocity dispersion. However, by applying the tensor virial theorem to small parts of a galaxy one may derive theoretical rotation curves and curves for the variation of line-of-sight velocity dispersion with radius, and hence show that the ratios of mean values figured here are good estimates of the relevant ratios of peak to central values (Binney 1980). This being so, the fact that the majority of the observational points plotted in these figures lie below the $\delta = 0$ curve in Fig. 1a and below the dashed median curve in Figure 1b indicates that the shapes of elliptical galaxies are not determined by rotational kinetic energy; rather their shapes are due to anisotropic velocity dispersions. In fact, if elliptical galaxies are nearly oblate, rotation does not even make an appreciable contribution to the shapes of those galaxies whose points in Figure 1a lie in the lowest region, where the curves of constant δ run almost vertically.

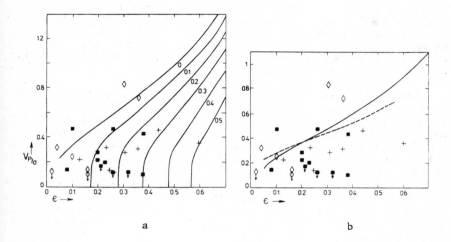

a   b

Fig. 1.   The relationship between (v/σ) and ε given by equation (11). 1a shows $(v_p/\sigma) \equiv \pi/4 \, (v/\overline{\sigma})$ for oblate spheroidal models having $\overline{\delta} = 0, 0.1,$ ... etc.   The factor of $\pi/4$ allows for the effect of projecting the true rotation velocity down a line of sight through the equatorial plane.
1b applies to prolate spheroidal galaxies having $\delta = q = 0$; the full curve shows the apparent (v/σ) and ε of such a galaxy when it is viewed down the short axis of the equatorial plane. Random orientation of the principal axes of these galaxies to the line of sight would cause half to lie above and half below the dashed line. Some observed values of (v_m/σ_c) are shown as crosses (Illingworth 1977), and squares (Schechter & Gunn 1979) and diamonds (Davies 1980).

However, it does not follow from Fig. 1a and 1b that the shapes of elliptical galaxies are not determined by the distribution of their stars with respect to angular momentum. Thus if these systems are oblate spheroids, one may in principle construct acceptable models in which the required excess of pressure in the equatorial plane comes about because the azimuthal velocity dispersion σ_ϕ is greater than the component σ_z in the z-direction. In such a model the stellar distribution function is a function of energy and the z-component of angular momentum only, and the radial velocity dispersion $\sigma_{\bar{\omega}} = \sigma_z$. The velocity ellipsoids are prolate spheroids that point in the azimuthal direction. However this class of models seems very unnatural and contrasts with the possibility that a non-classical 'third' integral plays the leading role. In this case $\sigma_\phi \simeq \sigma_{\bar{\omega}} \neq \sigma_z$, and the velocity ellipsoids are more nearly oblate spheroids whose shortest axes point out of the equatorial plane. Aarseth and I (Aarseth & Binney 1979) have shown that this situation arises naturally when a stellar system relaxes from a reasonably chaotic initial condition, whereas I do not know of a mechanism for manufacturing non-rotating galaxies whose distribution functions depend on energy and angular momentum alone. Furthermore, observations of stellar motions in the solar neighbourhood show that the galactic halo is not constructed on this plan.

Equation (9) enables one to show an interesting conclusion concerning the degree of velocity dispersion anisotropy at the galaxian centre and the orientation of the velocity ellipsoids there. I indicated above that from the point of view of the tensor virial theorem, rotation or pressure anisotropy are equivalent means of maintaining an aspherical shape. There is one respect, however, in which this is not entirely true. Near the centre any systematic rotation velocity must go to zero. This is consistent with constant ellipticity and zero pressure anisotropy in an oblate spheroidal galaxy because, if we subtract the z-diagonal equation of the set (9) from the corresponding x-diagonal equation, we have (neglecting T_{zz})

$$(S_{xx} - S_{zz}) + 2T_{xx} + (\Pi_{xx} - \Pi_{zz}) + (W_{xx} - W_{zz}) = 0 \qquad (12)$$

In the absence of pressure anisotropy the first and third terms in equation (12) vanish when we take the volume τ to be, say, that interior to a given isodensity surface of major axis length a. In the central near-homogeneous part of the galaxy $(W_{xx}-W_{zz}) \sim a^5$ and $T_{xx} \sim v_{rot}^2 a^3$, so $(v_{rot}^2)^{\frac{1}{2}} \sim a$ near the centre as expected. But if we consider a model that owes its shape to pressure anisotropy, we require $\Pi_{xx} - \Pi_{zz} \neq 0$ beyond the centre, with the result that a non-zero surface term $(S_{xx} - S_{zz})$ appears in the equation for the equilibrium of the innermost region. Furthermore, this surface term has the same sign as the gravitational term $(W_{xx} - W_{zz})$. Thus the central region of an anisotropy-pressure dominated elliptical galaxy is subject to two stresses that try to make it round; in addition to the gravitational stress represented by $(W_{xx} - W_{zz})$ it is being squeezed by the anisotropic pressure of the surrounding material through $(S_{xx} - S_{zz})$. And the latter term comes to dominate near the centre, because it goes to zero only as a^3 rather than as a^5. This in turn requires that $\Pi_{xx} - \Pi_{zz}$ go to zero only like a^3, and hence that the velocity dispersion anisotropy be <u>constant</u> near the centre. Furthermore, far from the galaxian centre it is likely that the long axes of the velocity ellipsoids point nearly to the galaxian centre, just as the velocity ellipsoid of the high-velocity stars in the solar neighbourhood point in the centre direction. But this situation cannot pertain close to the centre, because such an arrangement is incompatible with the existence of finite velocity dispersion anisotropy there. It is noteworthy that Schwarzschild's model shows precisely the reorientation of velocity ellipsoids near the centre which these considerations predict (Merritt 1980).

APPLICATIONS OF THE EQUATIONS OF STELLAR HYDRODYNAMICS

If one transforms equation (2) into a spherical coordinate system, one obtains an equation that enables one to discuss the steady-state equilibrium of a spherically-symmetric galaxy. We now know that elliptical galaxies do not have isotropic velocity dispersions, but if a galaxy is spherically-symmetric, it seems reasonable to assume that the velocity ellipsoid at any point is such that one principal axis points toward the galaxian centre and the other two are of equal length and point perpendicular to the latter direction. Thus dropping the term on the left hand side of equation (2) by the assumption of a steady state, and dotting through by the unit radial vector $\hat{r}$, one obtains

$$\frac{\partial}{\partial x_i}(\rho \hat{r}_k \overline{v_k v_i}) - \rho \overline{v_k v_i} \frac{\partial \hat{r}_k}{\partial x_i} = -\rho \hat{r}_k \frac{\partial \Phi}{\partial x_k} . \tag{13}$$

The first term on the left hand side of (13) is just the divergence of the vector $s_i = \hat{r}_k \overline{v_k v_i}$, so using the usual formula for the divergence in spherical coordinates, and exploiting the fact that neither s_θ nor s_ϕ can depend on θ or ϕ, one has

$$\frac{\partial}{\partial x_i}(\rho \hat{r}_k \overline{v_k v_i}) = \frac{1}{r^2} \frac{d}{dr} (r^2 \rho s_r) \equiv \frac{1}{r^2} \frac{d}{dr} (r^2 \rho \sigma_r^2), \tag{14}$$

where $\sigma_r^2 \equiv \hat{r}_k \overline{v_k v_i} \hat{r}_i$ is the r-diagonal component of the velocity dispersion tensor. Furthermore, $\hat{r}_k = x_k/|r|$, which enables one to evaluate the second term in (13), and the right hand side of that equation is nothing but $-\rho(d\Phi/dr)$. Thus (13) becomes

$$\frac{1}{r^2} \frac{d}{dr} (r^2 \rho \sigma_r^2) = \frac{\rho}{r} (\delta_{ki} - \frac{x_k x_i}{r^2}) \overline{v_k v_i} - \rho \frac{d\Phi}{dr}$$

$$= \frac{\rho}{r} (\sigma_r^2 + 2\sigma_t^2 - \sigma_r^2) - \rho \frac{d\Phi}{dr} . \tag{15}$$

Finally, if one introduces a parameter β by $\beta = 1 - \sigma_t^2/\sigma_r^2$, then expanding (15) one gets

$$\frac{1}{\rho} \frac{d}{dr} (\rho \sigma_r^2) + \frac{2\beta \sigma_r^2}{r} = -\frac{d\Phi}{dr} = -\frac{GM(r)}{r^2} \tag{16}$$

61

where M(r) is the total gravitating mass interior to r.

Equation (16) is the key to determining the mass distribution in elliptical galaxies, and it is useful to discuss the qualitative nature of its solutions. First note that the quantity ρ that appears in this equation is the mass density of some stellar population that has radial velocity dispersion σ_r; there is no <u>a priori</u> reason to believe that it is closely related to the gravitating mass density $\rho_t = (dM/dr)/(4\pi r^2)$ that is responsible for generating the potential Φ. Therefore, suppose that we observe ρ and σ_r for some species and let us ask what we can learn about the total mass density ρ_t. Introduce the logarithmic radius x = ln r and the circular velocity at r, $v_c(r) = (GM/r)^{\frac{1}{2}}$. Then equation (16) may be put into the form

$$\frac{d \ln \rho}{dx} = -\left[(\frac{v_c}{\sigma_r})^2 + \frac{d \ln \sigma_r^2}{dx} + 2\beta\right] \equiv -\alpha .\qquad(17)$$

This equation says: (a) The logarithmic luminosity gradient of an isothermal isotropic model is given by the square of the ratio of the circular velocity to the stellar velocity dispersion. Typically $\alpha \simeq 2.6$ for giant ellipticals, which implies $v_c \simeq 1.6\sigma$ for $\beta = 0$. If, as is likely, these galaxies have $\beta > 0.3$, then $v_c < 1.4 \sigma_r$. Notice that a direct measurement of v_c (from the dynamics of a gaseous disk, for example) enables one to determine β. (b) β must go to zero near the centre of a galaxy that has no point-like nucleus, because in this case both v_c and the two derivatives in equation (17) go to zero as $x \to -\infty$. Therefore $\beta \to 0$ as $x \to -\infty$ too. But depending on exactly how fast β goes to zero, it may still play a dominant role near the centre. If, as it often appears to be the case, the galaxy has an essentially stellar nucleus, there is no physical or mathematical reason why β should become small near the centre, and the effect of this term <u>must</u> be considered. Duncan and Wheeler (1980) have recently shown in relation to the nucleus of M87 that different assumptions regarding β can radically alter one's conclusion regarding M(r). (c) Sharp changes in the gradient of σ_r produce complex features in the luminosity profile, because the first and second terms on the right hand side of equation (17) produce opposite effects one after another:

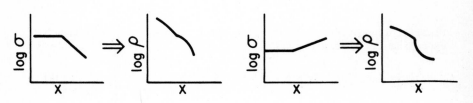

DYNAMICS OF ELLIPTICAL GALAXIES

It is tempting to consider the velocity-dispersion curve of an elliptical galaxy as analogous to the rotation curve of a spiral. Unfortunately no such analogy exists because, whereas the circular velocity at any radius pretty much determines the mass interior to that radius, the velocity dispersion at a given radius determines nothing at all. Indeed even when one adds to $\sigma_r(r)$ a knowledge of, say, the power-law index α (assumed constant) with which the luminosity density declines with increasing radius, one has to fit three parameters before one can recover the mass interior to a given radius: equation (17) regarded as an equation for σ_r^2 (α, β and v_c remaining constant) has solution

$$\sigma_r^2 = A \left(\frac{r}{r_o}\right)^{\alpha - 2\beta} + \frac{v_c^2}{\alpha - 2\beta} \quad . \tag{18}$$

All three of A, β and v_c have to be fitted by measuring σ_r at different radii, before $M(r) = v_c^2 r/G$ can be determined. If any of these quantities varies with r, one will have more unknowns than equations, and the problem has no unique solution.

Actually the situation regarding recovery of mass-data from luminosity and velocity dispersion data is even worse than this would indicate. For observed brightnesses and velocity dispersions do not relate to any one class of star as we have supposed above. Essentially all the light from an old stellar population comes from the very small fraction of stars that are on the giant and horizontal branches, and the mass-to-light ratio and line-strength characteristics of these stars can vary with radius, for example, on account of the steep metallicity gradients that are commonly observed at the centres of giant ellipticals. If $\beta = 0$ it is possible to employ equation (16) even when one observes different sets of unrepresentative stars at different radii provided one interprets ρ in a suitable way. But when $\beta \neq 0$ no such result obtains and the interpretation of stellar observations in terms of mass models becomes a risky business indeed:-

Divide the overall stellar population into classes $\alpha = 1$ .. N of near-identical stars; the mass-density of each class will satisfy the collisionless Boltzmann equation in the overall potential, and therefore the ρ_α, $\sigma_{r\alpha}$ and β_α values associated with each class will satisfy an equation like (16). If one defines $\ell_\alpha(r)$ to be the luminosity density associated with the class α, then obviously the ratio ℓ_α/ρ_α is independent of r, so that if one multiplies top and bottom of the first term of the αth equation like (16) by this ratio, one obtains

$$\frac{1}{\ell_\alpha} \frac{d}{dr} (\ell_\alpha \sigma_{r\alpha}^2) + \frac{2\beta_\alpha \sigma_{r\alpha}^2}{r} = -\frac{GM(r)}{r^2} \quad . \tag{19}$$

Now, if one supposes that all the $\beta_\alpha = 0$, then the overall velocity

63

dispersion σ^2 that one measures spectroscopically is the weighted mean of the velocity dispersions of the individual classes,

$$\sigma^2 = \sum_{\alpha} \ell_{\alpha} \sigma_{\alpha}^2 / \Sigma \ell_{\alpha} \quad , \tag{20}$$

and the total luminosity is $\ell = \Sigma \ell_{\alpha}$. Therefore if one multiplies each equation of the set (19) by ℓ_{α} and adds all the equations together, and then divides the result by ℓ, one obtains an equation,

$$\frac{1}{\ell} \frac{d}{dr} (\ell\sigma^2) = - \frac{GM(r)}{r^2} \quad , \tag{21}$$

that gives M(r) in terms of observable quantities. Notice, however, that the luminosity density ℓ that appears on the bottom of (21) must be composed of the same components ℓ_{α} that weight the velocity dispersions of the different classes. In practice this means that one should not use for ℓ the broad-band luminosity density that one obtains photographically, but one that is proportional to the number of photons by which the absorption lines, from which σ^2 is deduced, fall below the continuum at each radius.

If $\beta \neq 0$ life is not so simple. Indeed in this case ℓ_{α} times equation (19) is no longer linear in the observables, so one cannot simply add the equations and then divide by the total luminosity. Therefore in this case it becomes necessary to know the parameters ℓ_{α}, β_{α} etc. of all the populations that contribute significantly to the observed quantities at any radius if one wishes to derive a secure mass model from spectroscopic data. And again I must emphasize that observations of rotation in elliptical galaxies clearly imply that strong velocity anisotropy is present even at the very centres of elongated elliptical galaxies, so that it cannot be argued that the cores of ellipticals are appreciably relaxed and hence characterized by small values of β. Therefore this is a serious difficulty.

In principle a reliable route to M(r) is from observations of X-ray emission by hot gas trapped in the potential well of the galaxy, for the density ρ and temperature T of such gas obeys equation (16) with $\sigma_r^2 = RT/0.62$ and $\beta = 0$; this technique allows a complete solution of the analogous problem for rich clusters of galaxies (Strimpel & Binney 1979), and it has been applied to the outer regions of giant galaxies (Mathews 1978, Fabricant, Lecar & Gorenstein 1980). But unfortunately it is likely to be many years before we are able to observe the relatively cool ($T \simeq 10^6$K) gas which would enable us accurately to determine the central mass distributions of single galaxies. Probably a more promising line of attack on this problem is to obtain the rotation curves of the nuclear gaseous disks that are known to sit at the centres of many giant galaxies (Kotanyi & Ekers 1979).

As if this situation were not already bad enough, one may show that even if the mass-to-light ratio at the centres of ellipticals were constant, and the velocity dispersions there isotropic ($\beta = 0$), still there would be substantial uncertainty in the interpretation of the very compact nuclei that are often to be found at the very centres of giant elliptical galaxies, and indeed at the cores of large spheroidal components generally. Two of the best studied examples of this phenomenon in normal galaxies are the nucleus of M31 (Light, Danielson & Schwarzschild 1974) and that in NGC 3379 (de Vaucouleurs & Capaccioli 1979), and recently it has been suggested that black holes are responsible for cusps at the centres of M87 (Sargent et al. 1978) and of NGC 6251 (Young et al. 1979). However difficulty arises in interpreting observations of these systems on account of an unfortunate interplay between the effects of seeing and the non-linear nature of equation (16):-

Imagine that the total mass density $\ell(r)$ were made up of two near isothermal components

$$\rho(r) = \rho_a \left[1 + (r/a)^2 \right]^{-3/2} + \rho_1 \left[1 + r^2 \right]^{-3/2} . \qquad (22)$$

Furthermore, assume that the mass-to-light ratio of this material is constant, so that the surface brightness of the system is some constant multiple of the projected mass density. If we take $a < 1$ and $\rho_a = 3\rho_1/a$, so that exactly three quarters of the central surface brightness comes from the smaller component, then the overall central line of sight velocity dispersion σ_c will be dominated by the light of the central component, and will be effectively equal to the value

$$\sigma_c \simeq \sigma_a = \left(\frac{4\pi}{9} G\rho_a a^2 \right)^{\frac{1}{2}} \qquad (23)$$

that would characterize the central component if it were isolated rather than set inside the broader component. But on account of the relationship we have assumed between ρ_a and ρ_1, we have that

$$\sigma_c \simeq \left(\frac{4\pi}{9} G\rho_a a^2 \right)^{\frac{1}{2}} = \left(\frac{4\pi}{9} G\rho_1 \, 3a \right)^{\frac{1}{2}} . \qquad (24)$$

Therefore, as we shrink a below 1/3, keeping the central surface brightness constant, the central value of σ^2 drops below the value $\sigma_1^2 = (4\pi/9)G\rho_1$ characteristic of the broad component. Consequently the velocity dispersion profile of the entire system develops a deeper and deeper dip near $r = 0$ as a is diminished below 1/3. And it turns out that the de Vaucouleurs $r^{1/4}$ brightness profile is fairly well simulated by taking $a = 0.1$, which suggests that galaxies like NGC 3379 or M31, whose central brightness profiles can be well fitted by this law should have central depressions in σ. In the case of M31 such a depression in σ has been sought without success (Morton, Andereck & Bernard 1977), and in M87 σ actually rises towards the nucleus. Do these results necessarily rule out models in which the mass-to-light ratio is constant and the velocity

65

dispersion is isotropic? The answer to this question depends on whether a significant fraction of the light in which the nuclear velocity dispersion is based derives from the unresolved core of the nucleus. For it is in principle possible that a tight star cluster sits at the very centre of the galaxy and contributes a large portion of the unresolved central light. The velocity dispersion of this cluster can be made as large as one pleases simply by shrinking its (unresolved) core radius b whilst increasing its central density ρ_b in such a manner that the total light contributed by it to the unresolved region remains fixed; that is, the quantity $K(b_t/b)\rho_b b^3$ remains fixed, where $K(b_t/b)$ is a slowly-varying function of b and the tidal radius b_t of the hypothetical unresolved component. During this shrinking process, which will not be detectable photometrically, the velocity dispersion $\sigma \simeq ((4\pi/9)G\rho_b b^2)^{\frac{1}{2}}$ will rise as $b^{-\frac{1}{2}}$. Gurzadyan and Ozernoj (1980) have recently constructed a model for M87 on this plan, in which the increase in the observed velocity dispersion towards the centre of M87 is accounted for without resort to a super massive black hole.

REFERENCES

Aarseth, S.J. & Binney, J.J. (1979). Mon.Not.R.Astr.Soc., 185, 227.

Binney, J.J. (1978). Mon.Not.R.Astr.Soc., 183, 501.

Binney, J.J. (1980). Mon.Not.R.Astr.Soc., 190, 421.

Chandrasekhar, S. (1968). Ellipsoidal Figures of Equilibrium. New Haven: Yale University Press.

Davies, R.L. (1980). Mon.Not.R.Astr.Soc. in press.

Duncan, M.J. & Wheeler, J.C. (1980). Astrophys.J.Lett. 237, L27.

Fabricant, D., Lecar, M. & Gorenstein, P. (1980). Astrophys.J. submitted.

Gurzadyan, V.G. & Ozernoj, L.M. (1980). Pisma v Astronom. J. 6, 136.

Illingworth, I. (1977). Astrophys.J.Lett. 218, L43.

Kotanyi, C.G. & Ekers, R.D. (1979). Astron.Astrophys. 73, L1.

Light, E.S., Danielson, R.E. & Schwarzschild, M. (1974). Astrophys. J. 194, 257.

Mathews, W. (1978). Astrophys.J. 219, 413.

Merritt, D. (1980). Astrophys.J.Suppl. 43, in press.

Morton, D.C., Andereck, C.D. & Bernard, D.A. (1977). Astrophys.J. 212, 13.

Roberts, P.H. (1962). Astrophys.J. 136, 1108.

Sargent, W.L.W., Young, P.J., Boksenberg, A., Shortridge, K., Lynds, C.R. & Hartwick, F.D.A. (1978). Astrophys.J. 221, 731.

Schechter, P. & Gunn, J.E. (1979). Astrophys.J. 224, 472.

Strimpel, O. & Binney, J.J. (1979). Mon.Not.R.Astr.Soc. 188, 883.

Vandervoort, P. (1980). Preprint.

de Vaucouleurs, G. & Capaccioli, M. (1979). Astrophys.J.Suppl. 40, 699.

Young, P.J., Sargent, W.L.W., Kristian, J. & Westphal, J.A. (1979). Astrophys.J. 234, 76.

GALAXY MERGERS

Scott Tremaine

Institute for Advanced Study,
Princeton

1. INTRODUCTION

It is difficult to improve on Holmberg's (1940) introduction to this subject:

> The average space separation of extragalactic objects is rather small, compared with the dimensions of single objects. In a stationary universe we must expect a large number of encounters. Every close encounter between two objects will create large tidal disturbances, and the resulting loss of kinetic energy may be sufficient to effect a capture, i.e., to change the hyperbolic orbits of the objects into elliptical ones. Immediately after the capture, the elliptical orbits may be assumed to have rather large eccentricities. Every subsequent passage of a component through the pericenter of the relative orbit will, however, create new tidal effects and thus tend to decrease the eccentricity. The general result will be a gradual contraction of the relative orbit, which may continue until the two components form practically one object.

Holmberg (1941) also obtained quantitative estimates of the tidal energy loss in a close passage of two disk galaxies. He used an analog computer based on 74 movable lamps representing stars. The gravitational force on each star was determined from the net flux, which Holmberg measured by temporarily replacing the lamp with a photocell.

Despite Holmberg's pioneering work there was only a trickle of papers on mergers in the next thirty years. The main reason for the lack of interest was that close encounters and mergers seemed to be rare events. Suppose that any encounter with impact parameter less than R leads to a merger. The probability that a galaxy undergoes merger within a time T is

$$p = \pi R^2 \; <v_{rel}> \; NT, \tag{1}$$

where N is the number density of galaxies and $<v_{rel}>$ is the mean relative velocity. Let the Hubble constant be $H = 100h$ km s^{-1} Mpc^{-1}; then

$$p = 2 \times 10^{-4} \left(\frac{Nh^{-3}}{0.05 \text{ Mpc}^{-3}} \right) \left(\frac{Rh}{20 \text{ kpc}} \right)^2 \left(\frac{<v_{rel}>}{300 \text{ km s}^{-1}} \right) (HT) . \quad (2)$$

We see that if the capture radius R is identified with the visible size of galaxies, then the merger rate at present is very small. The probability p is independent of the Hubble constant since observations yield Nh^{-3} and Rh, and $HT \sim O(1)$. Thus, the conclusion that the merger probability is small was already reached by Holmberg (1940), even using a Hubble constant which was badly in error.

In the last decade three separate lines of argument have suggested that mergers are a common and important event in galactic evolution. In 1974 Ostriker, Peebles & Yahil argued that "there are reasons, increasing in number and quality, to believe that the masses of ordinary galaxies may have been underestimated by a factor of 10 or more." The evidence has continued to mount since then (see Faber & Gallagher 1979 for a review), and it now seems likely that most giant galaxies have massive halos which may extend 100 kpc or more in radius. In this case the capture radius R and merger probability p in (2) must be increased by large and uncertain amounts; moreover, the drag exerted by this halo on bound companion galaxies (see section 3) will lead to the orbital decay and merger of many bound systems.

Second, Toomre & Toomre (1972) and Toomre (1977) have argued that if galaxies are formed without large peculiar velocities, then it is natural to suppose that in many cases a galaxy and its nearest neighbor will form a bound pair. In the absence of tidal torquing, the galaxies will fall together and collide; if the collision is close enough to head-on then merging will occur. One argument that this scenario is common is that our Galaxy appears to form a bound pair with M31. Also, Aarseth & Fall (1980) found many mergers of loosely bound binaries in their cosmological N-body simulations.

The third and most striking argument is that the observed merger rate appears greatly to exceed the value given by equation (2). Toomre (1977) identifies ten pairs of interacting galaxies "in fairly advanced throes of merger" from ~ 4000 NGC galaxies. He argues that the tails by which he identifies his candidates cannot last much more than 5×10^8 yr; assuming a uniform rate throughout the past 10^{10} yr yields a merger probability $p = 0.05$. A reasonable extrapolation to higher rates in the past would yield an even higher value of p.

In this review I will discuss only mergers occurring after the process of galaxy formation is complete. Mergers between proto-galaxies may also be an important process (e.g. White & Rees 1978) but the details are less certain and will not be discussed here.

2. MERGERS OF GALAXIES OF COMPARABLE MASS

Consider a perturbing mass m_p which passes a galaxy at impact parameter p with a large velocity $\vec{v} = v\hat{z}$. The center of the galaxy is at the origin and the perturber's orbit is in the xz plane. For $|\vec{r}| \ll p$ the perturbing potential is

$$U(\vec{r}) = \frac{Gm_p}{r_p^3} \left[\frac{1}{2}r^2 - \frac{3}{2} \frac{(\vec{r}\cdot\vec{r}_p)^2}{r_p^2} \right] . \tag{3}$$

Thus, in the impulse approximation, a star at $\vec{r}$ receives a velocity increment

$$\Delta\vec{v} = - \int \nabla U dt = \frac{2Gm_p}{vp^2} (x,-y,0). \tag{4}$$

Assuming $\Delta\vec{v}$ is uncorrelated with $\vec{v}$ the change in energy per unit mass is

$$\Delta E = \frac{2G^2m_p^2}{v^2 p^4} (x^2 + y^2). \tag{5}$$

Averaging over a spherical galaxy of mass m_g gives

$$\Delta E = \frac{4G^2m_p^2m_g}{3v^2 p^4} <r^2> . \tag{6}$$

Thus, in the encounter of two identical galaxies, we get

$$\Delta E = \frac{8G^2m_g^3<r^2>}{3v^2 p^4} . \tag{7}$$

The orbital energy is $E = \frac{1}{4} m_g v^2$ so merging occurs if $\Delta E > E$ or

$$pv < \left[\frac{32}{3} G^2 m_g^2 <r^2> \right]^{1/4}. \tag{8}$$

The derivation above follows closely the argument of Spitzer (1958), who was concerned with the disruption of open clusters by interstellar clouds. Spitzer also recognized that the derivation failed when $v/p \lesssim \Omega$, where Ω is a typical angular velocity of a star in the galaxy or open cluster. In this limit the approximation that the stars are stationary during the encounter fails; for $v/p \ll \Omega$ the stars orbit many times during the encounter and, by adiabatic invariance, their energy change is small.

We can also treat another limit, the head-on collision at
large velocity. Consider a single star passing a spherical galaxy
with mass distribution $M(r)$. If the star has impact parameter p,
the perpendicular impulse which it receives is

$$\Delta v_\perp = \frac{2Gp}{v} \int_p^\infty \frac{drM(r)}{r^2 \sqrt{r^2 - p^2}} \quad . \tag{9}$$

Now if we let $\Sigma(r)$ be the surface density of the galaxy, and if
$M(r) = 2\pi \int_0^r \Sigma r dr$ is the projected mass inside a cylinder of
radius r, the formula above can be written as

$$\Delta v_\perp = \frac{2GM(p)}{pv} \quad . \tag{10}$$

If the star is regarded as a member of a second identical galaxy
colliding head-on, then the total change in energy in both galaxies
during the collision is

$$\begin{aligned}
\Delta E &= 2 \cdot 2\pi \int_0^\infty \tfrac{1}{2} \Delta v_\perp^2 \, \Sigma r dr \\
&= \frac{8}{3} \frac{G^2}{v^2} \int_0^\infty \left[\frac{M(p)}{p} \right]^3 dp .
\end{aligned} \tag{11}$$

Merger occurs if $\Delta E > E$ or

$$v^4 < \frac{32G^2}{3M} \int_0^\infty \left[\frac{M(p)}{p} \right]^3 dp . \tag{12}$$

For example, a modified Hubble law $\Sigma = \Sigma_0 (1 + r^2/a^2)^{-1}$ (Rood, Page,
Kintner & King 1972) has $M(r) = \pi \Sigma_0 a^2 \ln(1 + r^2/a^2)$ and $\Delta E = 298.2$
$G^2 \Sigma_0^3 a^4 / v^2$.

The head-on formula (12) complements the tidal formula (8),
and with a smooth interpolation between them they should yield
fairly accurate results for the energy loss in high-velocity
collisions. For low velocity collisions v and p may be replaced
by the velocity and distance at closest approach (where closest
approach is computed from the galaxy orbits on the assumption that
they are rigid, possibly interpenetrating spheres). The usefulness
and accuracy of the impulsive approximation have been stressed
primarily by Alladin and his coworkers (Alladin 1965, Sastry &
Alladin 1970, Sastry 1972, Alladin, Potdar & Sastry 1975); Toomre
(1977) and Dekel, Lecar & Shaham (1980) also point out that the
impulsive formulae are surprisingly accurate for low-velocity
collisions.

The simple capture criteria (8) and (12) can be checked by numerical simulations. However, even with our crude arguments the main uncertainty is in the observations because the galactic mass distribution is so poorly known (e.g. Faber & Gallagher 1979). Thus, the main purpose of numerical calculations is to determine the properties of the merger remnant (rotation speed, ellipticity, velocity dispersion, etc.).

There are two main approaches to numerical simulations of collisions and mergers of galaxies of comparable mass:

(1) Full N-body calculations have been carried out by Holmberg (1941), with 37 bodies per galaxy, by White (1978, 1979a) with 250 bodies per galaxy, by Roos & Norman (1979) with $\sim$ 30 bodies per galaxy, and by Dekel et al. (1980) with a 250 body galaxy perturbed by a single body of comparable total mass. White concentrates on mergers from bound or parabolic orbits while Dekel et al. examine hyperbolic collisions.

The disadvantages of N-body calculations are that statistical fluctuations are large because of the small number of bodies and that two-body relaxation effects cause evolution of the galaxies even when they are isolated (White 1978). For head-on collisions N rings can be used instead of N point masses (Toomre 1977).

(2) An alternative approach is to expand the potential in a complete set of harmonic functions and keep only the lowest orders of the expansion. The advantage of this method is that if K orders are kept and N particles are followed, only $\sim$ NK operations are needed per step, whereas an N-body code needs $\sim N^2$. Thus more particles can be used and statistical errors are reduced; however, all features on small scales are washed out. van Albada & van Gorkom (1977) studied the head-on collisions of two galaxies with N = 1000 particles per galaxy; because the system was axially symmetric they expanded the potential using Legendre polynomials up to index $\ell = 4$. Villumsen (1980) used N = 600 particles per galaxy and expanded the potential in tesseral harmonics up to $\ell = 4$. Miller & Smith (1980) use N = 50,000 particles per galaxy, work within a cube of side L, and keep the lowest $(64)^3$ Fourier coefficients of the potential.

A useful trick introduced by van Albada & van Gorkom is to use two separate coordinate systems in the potential calculations, one centered on each galaxy. This reduces the higher-order potential components during the initial approach.

Some of the results from these calculations include:

(1) The cross-section for mergers between rotating galaxies is strongly enhanced if the spin and orbital angular momenta are

aligned, and sharply reduced if the spins are opposite to the orbital angular momentum (White 1979a). This is not surprising since the velocity of the perturber relative to a prograde star is small so the perturbation acts for a longer time.

(2) The merger remnant usually has both a higher central density and a more extended outer envelope than its progenitors (e.g. Dekel et al. (1980) find that the radius containing half the mass shrinks but the radius containing 90% of the mass grows).

(3) Escaping stars carry away relatively little mass or energy, although they may carry away significant amounts of angular momentum in off-center collisions.

(4) Head-on collisions lead to prolate galaxies elongated along the line of the initial trajectory; off-center collisions lead to oblate galaxies flattened along the initial orbital plane.

All of the work done so far has been on mergers of spheroidal systems. These calculations are mainly useful for determining the properties of merger remnants at the centers of rich clusters (see section 4). It is at least as interesting to try to understand the merging of two disk systems, since spiral-spiral mergers are likely to be much more common than elliptical-elliptical mergers in the field. However, there has been relatively little progress in this area because, among other reasons, it is difficult to construct isolated stable disk systems, and the number of parameters to be studied is so large.

3. MERGERS OF SATELLITE GALAXIES

3.1 Dynamical friction

In the impulsive calculations of the previous section the energy change due to a perturber of mass m_p is $\propto m_p^2$. Thus the frictional effects which lead to mergers are due to second-order perturbations. Since the orbital kinetic energy is $\propto m_p$, a sufficiently low mass perturber will not be captured from an unbound orbit. However, a satellite galaxy of mass m_s in a bound orbit around a galaxy of mass $m_g \gg m_s$ may still merge because its orbital energy is lost gradually over many orbits.

This process may be regarded as a manifestation of dynamical friction (Chandrasekhar 1960). A galaxy of mass m_s moving at velocity $\vec{v}$ through an infinite homogeneous background of stars with density ρ and a Gaussian velocity distribution with one-dimensional dispersion σ suffers a drag

$$\frac{dv}{dt} = - 4\pi G^2 m_s \rho v^{-2} \left[\phi(x) - x\phi'(x)\right] \ \ln \Lambda \ , \tag{13}$$

where $x = 2^{-\frac{1}{2}}v/\sigma$, ϕ is the error function and $\Lambda = p_{max}/p_{min}$, where p_{max} and p_{min} are the maximum and minimum impact parameters considered. Usually $p_{min} = max(r_s, Gm_s/v^2)$, where r_s is the size of the satellite, and p_{max} is the scale size of the background. Notice that dynamical friction is a second-order effect in m_s since the force on m_s is $\propto m_s^2$.

One example of the use of equation (13) is Tremaine's (1976) estimate of the decay rate of the orbit of the Magellanic Clouds in the halo of our Galaxy. Alar Toomre has pointed out that Cox (1972) also treated the decay of a satellite orbit in a spherical galaxy using equation (13).

The frequency of mergers of satellite galaxies is relatively small unless the central galaxies have extended massive halos. If the halo mass distribution is an isothermal sphere with a very small core radius, then $\rho(r) \simeq \sigma^2/(2\pi Gr^2)$, and a satellite in a circular orbit has $v = \sqrt{2}\sigma$. From equation (13) the orbital radius evolves as

$$r^2(t) = r^2(0) - 0.605 \frac{Gm_s t}{\sigma} \ln \Lambda, \qquad (14)$$

or

$$r(t) = \left[r^2(0) - (52\text{kpc})^2(t/10^{10}\text{yr})(m_s/10^{10}M_\odot)(100\text{km s}^{-1}/\sigma)\ln \Lambda\right]^{\frac{1}{2}}. \qquad (15)$$

Now suppose that initially the number density of satellites is $n(r)$. The flux through a given radius is $\propto r^2 n(r) dr/dt \propto rn(r)$. Thus if $n(r) \propto r^{-\gamma}$ we expect depletion of bright (i.e. massive) satellite galaxies at small radii for $\gamma > 1$ and an overabundance of bright close satellites for $\gamma < 1$. In fact $\gamma \simeq 1.8$ (Peebles 1980) so there should be depletion; however, the amount of depletion is uncertain and small because tidal stripping may reduce m_s and thus $|dr/dt|$ as the galaxy spirals in. Ostriker & Turner (1979) point out that depletion may also be masked by brightening of incoming galaxies due to tidal shocks which induce star formation. The observational evidence for depletion has been discussed by Ostriker & Turner (1979), White & Valdes (1980), and White (1980). The interpretation of the observations is uncertain but there is no strong evidence for or against the rapid decay rate and short lifetimes of nearby satellites predicted by the dynamical friction formula if massive halos are present.

How much has a typical galaxy eaten? Peebles (1980) writes the number density of galaxies of mass m_s at separation r from a given galaxy as $n(r,m_s)dm_s = n_0(m_s)dm_s (r/r_0)^{-\gamma}$ where $n_0(m_s)dm_s$ is

73

the field density, $\gamma \cong 1.8$, and $r_0 \cong 3h^{-1}$Mpc (Davis, Geller & Huchra 1978). This expression is valid for $r \ll r_0$. We obtain the field density from Schechter's (1976) luminosity function, assuming constant M/L: $n_0(m_s)dm_s = n*(m_s/m*)^{-\alpha} \exp(-m_s/m*)dm_s/m*$, where $n* = 0.04 \, h^3 \, Mpc^{-3}$, $m* = 6.8 \times 10^9 \, h^{-2} (M/L)M_\odot$, $\alpha = 1.25$. Using equation (14) to determine $r(0)$ for $r(t) = 0$, we find that the accreted mass is

$$m_{acc}/m* = \int_0^\infty n_0(m_s)(m_s/m*)dm_s \int_0^{r(0)} 4\pi(r/r_0)^{-\gamma} r^2 dr$$

$$= 0.06 \left[(H_0 t) \, \ln \Lambda \quad (M/hL)(100 \text{ km s}^{-1}/\sigma) \right]^{0.6}. \tag{16}$$

For the luminous central parts of galaxies $M/hL \approx 15$ (Faber & Gallagher 1979), and if we adopt $\ln \Lambda \approx 2$, $H_0 t \approx 1$, then $m_{acc}/m* \approx 0.5 \, (100 \text{ km s}^{-1}/\sigma)^{0.6}$ independent of h. Within the large uncertainties a typical giant galaxy has eaten somewhat less than its own mass.

3.2 The validity of the dynamical friction formula

The dynamical friction formula (13) is appealing because of its simplicity. However, it is not rigorous; in particular, the formula is difficult to generalize from an infinite homogeneous background to a spherical system. To see this, consider (for example) a star in a galaxy which is initially on a circular orbit in the plane of the satellite orbit. Suppose the satellite orbit is also circular with radius r_s. The potential of the central galaxy may be written $U(r)$, and the angular speed of an object in a circular orbit of radius r is Ω, where $\Omega^2 = r^{-1}dU/dr$. The potential from the satellite may be written as a series of terms of the form $\phi_m(r) \cos m(\theta - \Omega_s t)$, where $\Omega_s = \Omega(r_s)$, m is an integer, and θ is the azimuthal angle in the orbital plane. Then the equations of motion of the star due to a single term in the perturbing potential are

$$\frac{d^2 r}{dt^2} - \frac{J^2}{r^3} = -\frac{dU}{dr} - \frac{d\phi_m}{dr} \cos m(\theta - \Omega_s t), \quad \frac{dJ}{dt} = + m\phi_m \sin m(\theta - \Omega_s t)$$

$$\tag{17}$$

where $J = r^2 d\phi/dt$ is the angular momentum. If the perturbation is turned on at $t = 0$ then to first order in ϕ_m the solution is

$$J_1 = -\left(\frac{\phi_m}{\Omega - \Omega_s}\right)_{r_0} \{\cos[m(\Omega_0 - \Omega_s)t + m\theta_0] - \cos m\theta_0\} ,$$

74

$$r_1 = - \frac{\left[d\phi_m/dr + 2\Omega\phi_m/r(\Omega-\Omega_s)\right]_{r_0}}{\kappa_0^2 - m^2(\Omega_0-\Omega_s)^2} \times$$

$$\{\cos\left[m(\Omega_0-\Omega_s)t + m\theta_0\right] - \cos\kappa_0 t\cos m\theta_0 + \frac{m(\Omega_0-\Omega_s)}{\kappa_0} \sin \kappa_0 t \sin m\theta_0\}$$

$$+ \frac{2\Omega_0\phi_m(r_0)}{r_0\kappa_0^2(\Omega_0-\Omega_s)} \cos m\theta_0(1-\cos\kappa_0 t) \ , \tag{18}$$

where the star is at (r_0, θ_0) at $t = 0$, $\Omega_0 = \Omega(r_0)$ and the epicyclic frequency κ_0 is given by $\kappa_0^2 = (d^2U/dr^2 + 3r^{-1}dU/dr)_{r_0}$. There is no secular torque on the star to first order. The calculation can also be carried to second order and there is still no secular torque. There is also no secular increase in the star's energy E since $dE/dt = \Omega_s dJ/dt$ by Jacobi's integral, and consequently no drag on the satellite analogous to dynamical friction, which is also second order in the perturbing potential. A similar conclusion was reached by Kalnajs (1972) in a more artificial but exactly soluble system.

There are two aspects to the resolution of this problem. First, consider only relatively close encounters (say, by setting P_{max} in the Chandrasekhar formula equal to $2p_{min}$). For these encounters the approximation of an infinite homogeneous background is valid since the impact parameter is much less than the scale size and a given star is unlikely to have more than one close encounter. However, this restriction decreases Chandrasekhar's drag force only by a factor $\sim \ln \Lambda$. Thus, a conservative estimate of the drag force is given by considering only close encounters, i.e. by formula (13) with $\ln \Lambda \sim 1$. The difference of a factor of $\ln \Lambda$ is not generally large enough to have observable consequences.

Second, and more important, the first order perturbations J_1 and r_1 in equation (18) diverge at resonances where $\Omega = \Omega_s$ (co-rotation resonance) and $\kappa_0^2 = m^2(\Omega_0 - \Omega_s)^2$ (Lindblad resonance). Thus, perturbations near these resonances are large. Moreover, the formula for the second order torque dJ_2/dt contains periodic terms of long period near these resonances. Let us restrict ourselves for the moment to the vicinity of the Lindblad resonance; the corotation resonance can be handled by similar techniques. The second-order torque near the Lindblad resonance at $\kappa = m(\Omega_0 - \Omega_s)$ due to divergent long-period terms is

$$\frac{dJ_2}{dt} = - \frac{m}{4\kappa_L r_L^2}\left(r \frac{d\phi_m}{dr} + \frac{2m\Omega}{\kappa} \phi_m\right)^2_{r_L} \frac{\sin \Delta t}{\Delta} \tag{19}$$

where $\Delta = \kappa - m(\Omega_0-\Omega_s)$. The resonance radius r_L is defined by

75

$\Delta(r_L) = 0$ and we have dropped the distinction between r_L and r_0 except in Δ. The torque on a given star (fixed Δ) grows like t until it drops out of resonance at $|\Delta t| \sim \pi$. As time goes on the number of stars in resonance decreases like t^{-1} but each one feels a torque $\propto t$. Thus the total torque is independent of time. The existence of secular torques at resonances was recognized by Lynden-Bell & Kalnajs (1972), who also derived a more general form of (19) for resonances in an arbitrary flat axisymmetric system. Eventually equation (19) fails because the perturbations on the stars which are still in resonance become non-linear. However, evolution of the satellite orbit will generally bring fresh stars into resonance so that a secular torque continues to be present.

In a real galaxy, with eccentric satellite and star orbits, the resonance structure is much more complicated than a single Lindblad resonance. The purpose of the simple example presented here is to show that near-resonant stars in a galaxy can exert secular torques on a satellite. These torques are the analogs, for spherical or axisymmetric systems, of the drag force described by Chandrasekhar's dynamical friction formula for infinite homogeneous systems. (To see this consider a resonance with azimuthal wavenumber m of order unity. Then $\phi_m \sim Gm_s/r$, where m_s is the satellite mass, and r is the orbital radius. We may set $\kappa \sim \Omega$, and the number of field stars in an interval $d\Delta$ is $\sim \rho r^2 d\Delta/\Omega$ where ρ is the mean galaxy density. Then integrating (19) over Δ we have $dJ_2/dt \sim G^2 m_s^2 \rho/\Omega^2 r$. The classical formula (13) yields the same result, $dJ_2/dt \sim m_s r dv/dt \sim G^2 m_s^2 \rho r v^{-2} \sim G^2 m_s^2 \rho/\Omega^2 r$, using $v \sim \Omega r$.) In principle it is possible to compute exactly the frictional force on a satellite in a specified orbit in a specified axisymmetric galaxy.

There are at least two situations where this kind of calculation will yield important new results. First, we can determine how fast frictional forces fall off beyond the outer edge of a galaxy; if the galactic halo ends at 50 kpc, what is the fate of a satellite in orbit at 100 kpc? Second, we can determine the orbital evolution of close companions of spiral galaxies which are influenced by the disk component of the central galaxy. For example, the disk may slow or halt merging of companions in prograde orbits by adding angular momentum to the orbit as fast as the halo removes it.

3.3 Numerical techniques

Numerical calculations of the merging of bound companions can be done using the methods described in the previous section. These techniques are best suited for systems of two galaxies of comparable mass although White's (1978) N-body experiments include a test of Chandrasekhar's dynamical friction formula for a satellite with 0.1 times the mass of the main galaxy; the formula works quite well.

If the mass of the satellite is much less than the mass of the central galaxy, then many dynamical times are needed for merger to occur. In this case N-body experiments are both expensive and subject to numerical error and two-body relaxation effects. An alternative approach has been proposed independently by Borne (1980) and by Lin & Tremaine (1980). They investigate a system containing three types of objects: (1) a central galaxy described by a potential $U(r)$; Lin & Tremaine use a point mass potential $U(r) = -GM/r$ while Borne uses a more realistic potential, (2) N stars of mass m which orbit the central galaxy, (3) a satellite galaxy of mass m_s, also orbiting the central galaxy. The stars interact gravitationally with the central galaxy and the satellite; the difference from a standard N-body program is that the stars do not interact with each other. Turning off the star-star attraction in this way eliminates two-body relaxation, increases numerical accuracy and greatly increases computational speed ($\sim$ N calculations per step instead of N^2). The only sacrifice is that the self-gravity of the outer parts of the central galaxy has been neglected.

I will describe two sample results using this technique. To investigate frictional effects on a satellite orbiting beyond the outer edge of a galaxy, Lin and I constructed a galaxy with central mass M = 1 (in units with G = 1) surrounded by a halo of N = 450 stars with total mass Nm = 1. The stars were initially distributed with a phase space density f depending only on energy $\varepsilon = M/r - \frac{1}{2}v^2$. We chose $f(\varepsilon) \propto \varepsilon^{2.5}$ for $0.5 < \varepsilon < 2.5$ and $f(\varepsilon) = 0$ for $\varepsilon < 0.5$ or $\varepsilon > 2.5$. Thus the maximum radius which a star could reach in its unperturbed orbit was r = 2; however, over 95% of the stars have r < 1 and we will call $r = r_0 = 1$ the "edge" of the halo. The satellite galaxy mass was $m_s = 0.1$; it was placed in a circular orbit of radius $r_s = 2.28$, over a factor of two larger than the halo edge at r_0. Its initial orbital period was $2^{\frac{1}{2}}\pi \cdot (2.28)^{3/2} = 15.30$ units. (Recall that m_s is attracted by an effective mass M+Nm = 2, since it interacts with both the stars and the central mass.) The evolution of the angular momentum J_s of the satellite is shown in Figure 1. The angular momentum changes slowly at first but with increasing speed as the orbit decays. An arrow marks the point corresponding to a circular orbit at the edge of the halo $r_0 = 1$; notice that the orbital decay is very fast beyond this point as the satellite is within the halo itself. These results show that strong frictional effects are present even in a satellite orbiting well outside the radius of most of the stars.

Our second experiment began with the same initial conditions, but the satellite galaxy was frozen into its initial circular orbit at $r_s = 2.28$ for 120 time units ($\sim$ 8 orbital times) before it was permitted to decay. An initial freezing period may correspond more closely to the way the satellite galaxy was actually formed. In this case the decay time for the satellite to reach the edge of the halo was 138 units after release, whereas in the first experiment the decay time was only 51 units. The late stages of the decay are very similar in both cases, suggesting that the system loses all memory of the initial 'freezing' once the decay has begun.

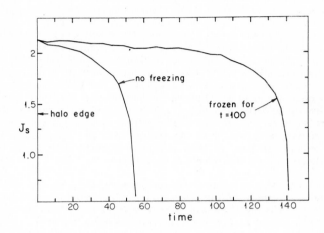

Fig. 1. Decay of the orbital angular momentum J_s of a satellite of mass $m_s = 0.1$ in orbit around a point mass $M = 1$ surrounded by a halo of $N = 450$ stars of mass $m = 1/450$. In one case the satellite was frozen into its initial circular orbit for 120 time steps before it was released.

4. MERGERS IN RICH CLUSTERS

The one-dimensional velocity dispersion in a typical rich cluster is $\sigma_{cl} \sim 1000$ km s^{-1}. Thus encounters between cluster galaxies usually take place at high relative velocity and mergers do not occur. However, mergers can occur by a slightly more subtle process. Dynamical friction causes the orbits of cluster galaxies to decay, just like the orbits of satellite galaxies in a halo (Lecar 1975). From equation (15), substituting $\ln \Lambda \sim 1$ (a lower limit, see section 3) and $\sigma \sim \sigma_{cl}$, we see that the orbit of a typical giant galaxy with $m_s \sim 10^{12} M_\odot$ will decay all the way to the cluster center in 10^{10} yr if $r(0) \lesssim 150$ kpc. As galaxies accumulate in the cluster center they merge by the processes we have already discussed. Thus a single large merger remnant may be built up (Ostriker & Tremaine 1975, White 1976).

There is strong evidence that these remnants correspond to the cD galaxies identified by Morgan and his colleagues by morphological

examination of photographic plates. These are defined by Matthews,
Morgan & Schmidt (1964) and Morgan, Kayser & White (1975) as supergiant galaxies having an elliptical-like nucleus surrounded by an
extensive envelope. Most known cD's are in rich clusters though
some are found in groups (Morgan et al. 1975). It is attractive
to identify the cD's in rich clusters as merger remnants because:

(1) The cD is often much more luminous than any other galaxy in
the cluster (e.g. Dressler 1978). This statement is not very
precise as it stands since cD's are already defined to be very
luminous ('supergiant'). More accurately, we can say that the
brightest members of clusters of galaxies are often more luminous
than we would expect from the statistics of the luminosity function
and that these exceptionally bright galaxies usually have extended
halos (Oemler 1976, Tremaine & Richstone 1977, Dressler 1979).
Both characteristics are necessary for a cD, and both are consistent
with cD's being merger products: the observed luminosities are
roughly equal to the total luminosity of the galaxies which spiral
to the center in a Hubble time (White 1976) and the extended halos
are observed to form in N-body simulations of mergers (cf. section 2).

(2) The cD galaxy is usually found near the cluster center (Morgan
& Lesh 1965, Leir & van den Bergh 1977, R.A. White 1978), consistent
with the buildup of a remnant at the bottom of the cluster's potential well.

(3) Many cD's have double or multiple nuclei (Morgan & Lesh 1965).
A good example is NGC 6166, the cD in A2199 (Minkowski 1961).
These may be tidally stripped remnants of galaxies which have
spiralled into the cD. Hoessel (1980) finds that the fraction of
first-ranked cluster galaxies with multiple nuclei separated by
$< 10h^{-1}$ kpc (about 1 in 4 from a sample of 100 galaxies) is consistent with theoretical estimates of merger rates. Similarly,
Rood & Leir (1979) point out that in $\sim 25\%$ of Bautz-Morgan Type I
clusters (clusters containing a central cD, Bautz & Morgan 1970),
the first-ranked galaxy is part of a 'dumb-bell' system containing
two galaxies differing by ≤ 1 magnitude. Contrary to Rood & Leir's
conclusion, I believe that the observed fraction is roughly consistent with theoretical merger rates. The mean spatial separation
of Rood & Leir's pairs is $\sim 20h^{-1}$kpc. From Jenner (1974) and
Dressler (1979) the velocity dispersion in a typical cD envelope
is ~ 500 km s^{-1}, corresponding to a circular velocity of ~ 700 km s^{-1}.
The resulting orbital time is $\sim 1.8 \times 10^8$ h^{-1}yr. The decay time for
equal mass galaxies is about one orbital time; since the second
galaxy is perhaps a factor of two fainter than the cD, a rough
estimate for the decay time is 4×10^8 h^{-1}yr. Since 25% of the cD's
are in dumbbells we conclude that a dumbbell is formed about every
(1 to 2)$\times 10^9$ h^{-1}yr. For want of better information we take the
typical cluster age to be $\frac{1}{2}H_0^{-1} = 5 \times 10^9$ h^{-1}yr; thus Rood & Leir's
Bautz-Morgan Type I clusters have gone through 2-5 dumbbell stages
on average. This is roughly the rate predicted by merger simulations

(Hausman & Ostriker 1978), and is also roughly the rate required to produce a cD galaxy one or two magnitudes brighter than the other galaxies in the cluster.

(4) Galaxies get fat when they eat. Ostriker & Hausman (1977) and Hausman & Ostriker (1978) have stressed that this effect should be detectable as a correlation between the luminosity of the first ranked galaxy and the structure parameter $\alpha = (d\ln L/d\ln R)_{R_0}$, where L is the luminosity inside an aperture of radius R. Hoessel (1980) has measured α for 90 first ranked galaxies at $R_0 = 10 \ h^{-1}$ kpc and finds a strong correlation which follows the predictions of Hausman & Ostriker.

(5) The major axes of cD galaxies are aligned with the major axes of their clusters (Sastry 1968, Carter & Metcalfe 1980). This alignment is natural if the cD has grown by accreting cluster members.

Many further details of all these tests are given in the references. In the last five years an impressive array of evidence has been accumulated in support of the hypothesis that many first ranked cluster galaxies (and possibly all cD galaxies) are merger remnants. We must remember, however, that the center of a rich cluster is a complex environment, and many other processes may play a role in the evolution of galaxies there (e.g. Cowie & Binney 1977).

5. DO MERGED SPIRALS MAKE ELLIPTICALS?

Toomre & Toomre (1972) and Toomre (1977) have speculated that most or all elliptical galaxies may be the remnants of merged spirals. The original motivation for this suggestion was the calculation described in section 1, which showed that the observed merger rate (times a factor of three extrapolation to account for a higher rate in the past) could produce the observed fraction of ellipticals ($\sim 15\%$) in a Hubble time.

Since elliptical galaxies have virtually no interstellar gas, the merger process must remove $\sim 10^{10} M_\odot$ of gas in the collision of two normal spiral galaxies. This does not appear to present serious difficulties. The most plausible mechanism is that a burst of star formation driven by tidal or collisional shocks produces enough supernovae to drive a wind which expels the gas from the merger remnant. The peculiar colors of interacting galaxies suggest that star formation bursts do occur in galaxy collisions (Larson & Tinsley 1978). The average supernova rate in our Galaxy already supplies a large fraction of the energy input to the interstellar medium, so it is plausible that a burst of supernovae could sweep the gas out.

A second potential problem was raised by White (1979b), who pointed out that if mergers occur between unbound systems of comparable mass, then the merged remnant may have more angular momentum than observed elliptical galaxies. This is not a serious problem since most mergers occur between initially bound systems, as described in section 1 and mentioned by White. In fact, N-body simulations by Jones & Efstathiou (1979) and Aarseth & Fall (1980) yield quantitative agreement with observations. The latter authors expressed their results in terms of the dimensionless number $\lambda = J|E|^{1/2} G^{-1} M^{-5/2}$, where J, E and M are the angular momentum, energy and mass of the merged remnant. They found $<\lambda> = 0.07 \pm 0.02$, and estimated $<\lambda> \approx 0.08$ from observations, in good agreement. This result is encouraging but does not provide a strong test of the merger hypothesis since tidal torques also yield $<\lambda> = 0.07 \pm 0.03$ (Efstathiou & Jones 1979). A more stringent test is to ask whether mergers produce the observed relation between flattening and V/σ (the ratio of rotation speed to velocity dispersion). This test requires accurate numerical experiments on the merging of disk systems.

However, there are several serious problems with the hypothesis that ellipticals are merged spirals:

(1) Ostriker (1980) has pointed out that the fractional abundance of ellipticals in the centers of cD clusters is high (e.g. Oemler 1974), but, as pointed out in section 4, mergers cannot occur in this environment because the encounter velocities are too high to lead to capture (cf. equation 12). Merging can only occur during the collapse of the cluster, when the velocity dispersion is low, but at this stage the density enhancement is also small so one would not expect an enhanced merger rate. This argument can and should be checked by N-body simulations since it is possible that merging may occur in sub-clumps during the collapse; existing simulations (Aarseth & Fall 1980) are inconclusive because they do not produce rich clusters (structures with velocity dispersion $\sim$ 1000 km s^{-1}).

(2) Ostriker (1980) also points out that ellipticals satisfy both a color-luminosity relation (Visvanathan & Sandage 1977) and a metallicity-luminosity relation (Faber 1977) while spirals do not. This is difficult to explain if ellipticals are made from merged spirals.

(3) Dwarf elliptical satellite galaxies cannot form by mergers with other satellites since their relative velocities are too high. Also, it is difficult to imagine what they merged from: there are $\sim$ 10 dwarf ellipticals within a few hundred kpc of our Galaxy and M31 and no spiral or irregular galaxies except for the Magellanic Clouds and M33, which are much more massive. Moreover, the properties of dwarf ellipticals (metallicity, number density, radius, etc.) appear to join smoothly onto those of giant ellipticals,

which suggests a common formation process.

(4) Elliptical galaxies have more globular clusters per unit luminosity than spirals, by a factor of $\sim$ 3 (Harris & Racine 1979). However, the number of globulars per unit of luminosity in the spheroidal component is about the same for both types, as might be expected since the globulars and the spheroid are both Population II. This result is not expected in the merger picture, since in a merger the light of the disk is added to the spheroidal luminosity while the number of globulars remains the same. The problem could in principle be reduced if new globulars were produced in the merger process in the right amount, but there is no evidence that this process has occurred.

Let me make one final objection (though I will not dignify it with a number). In many ways the bulges of spiral galaxies look very similar to ellipticals (surface brightness distribution, metallicity, ellipticity, velocity dispersion, etc.). Occam's razor suggests that both should be made the same way, and spiral bulges are certainly not made by merging spirals.

Many of the arguments in this section can be sharpened (or dulled) considerably by more careful thought. At the moment there seem to be grave doubts that most ellipticals can be made from merged spirals. However, there is strong evidence, described in section 1, that several hundred NGC galaxies have undergone mergers in the past. Then where are they, and what do they look like? Are some ellipticals merger remnants? If so, can they be recognized in any way? And if not, as Toomre (1977) remarked, "Where else have they possibly gone?"

6. CONCLUSIONS

The work on mergers which I have described is perhaps most important as part of a fundamental change in our conception of galaxies. Over the last few years we have finally discarded the idea of galaxies as "island universes" which are born and die in splendid isolation. The replacement is a richer and more complex picture, only partly drawn, which (we hope) will lead to a marked improvement in our understanding of the observations.

I have benefited enormously from conversations with S.M. Fall, J.P. Ostriker and Alar Toomre. I am grateful to John Bahcall, James Binney, and Herb Rood for comments on a draft manuscript. This work was supported in part by NSF Grant PHY 79-19884.

REFERENCES

Aarseth, S.J. & Fall, S.M. (1980). Astrophys.J., 236, 43.
van Albada, T.S. & van Gorkom, J.H. (1977). Astron.Astrophys., 54, 121.
Alladin, S.M. (1965). Astrophys.J., 141, 768.
Alladin, S.M., Potdar, A. & Sastry, K. (1975). In Dynamics of Stellar Systems, ed. A. Hayli, p. 167. Reidel, Dordrecht.
Bautz, L. & Morgan, W.W. (1970). Astrophys.J.Lett., 162, L149.
Borne, K. (1980). Ph.D. Thesis, California Institute of Technology.
Carter, D. & Metcalfe, N. (1980). Mon.Not.R.Astr.Soc., 191, 325.
Chandrasekhar, S. (1960). Principles of Stellar Dynamics. Dover, New York.
Cowie, L. & Binney, J. (1977). Astrophys.J., 215, 723.
Cox, L.P. (1972). Unpublished.
Davis, M., Geller, M.J. & Huchra, J. (1978). Astrophys.J., 221, 1.
Dekel, A., Lecar, M. & Shaham, J. (1980). Astrophys.J., in press.
Dressler, A. (1978). Astrophys.J., 222, 23.
Dressler, A. (1979). Astrophys.J., 231, 659.
Efstathiou, G. & Jones, B.J.T. (1979). Mon.Not.R.Astr.Soc., 186, 133.
Faber, S.M. (1977). In The Evolution of Galaxies and Stellar Populations, ed. B.M. Tinsley and R.B. Larson, p. 157. Yale University Press.
Faber, S.M. & Gallagher, J.S. (1979). Ann.Rev.Astron.Astrophys., 17, 135.
Harris, W. & Racine, R. (1979). Ann.Rev.Astron.Astrophys., 17, 241.
Hausman, M.A. & Ostriker, J.P. (1978). Astrophys.J., 224, 320.
Hoessel, J.G. (1980). Astrophys.J., in press.
Holmberg, E. (1940). Astrophys.J., 92, 200.
Holmberg, E. (1941). Astrophys.J., 94, 385.
Jenner, D. (1974). Astrophys.J., 191, 55.
Jones, B.J.T. & Efstathiou, G. (1979). Mon.Not.R.Astr.Soc., 189, 27.
Kalnajs, A. (1972). In Gravitational N-Body Problem, ed. M.Lecar, p. 13. Reidel, Dordrecht.
Larson, R. & Tinsley, B. (1978). Astrophys.J., 219, 46.
Lecar, M. (1975). In Dynamics of Stellar Systems, ed. A. Hayli, p. 161. Reidel, Dordrecht.
Leir, A. & Van den Bergh, S. (1977). Astrophys.J.Suppl., 34, 381.
Lin, D.N.C. & Tremaine, S. (1980). In preparation.
Lynden-Bell, D. & Kalnajs, A.J. (1972). Mon.Not.R.Astr.Soc., 157, 1.
Matthews, T.A., Morgan, W.W. & Schmidt, M. (1964). Astrophys.J., 140, 35.
Miller, R.H. & Smith, B.F. (1980). Astrophys.J., 235, 421.
Minkowski, R. (1961). Astron.J., 60, 558.
Morgan, W.W., Kayser, S. & White, R. (1975). Astrophys.J., 199, 545.
Morgan, W.W. & Lesh, J. (1965). Astrophys.J., 142, 1364.
Oemler, A. (1974). Astrophys.J., 194, 1.
Oemler, A. (1976). Astrophys.J., 209, 693.

Ostriker, J.P. (1980). Comm.Astrophys., 8, 177.
Ostriker, J.P. & Hausman, M.A. (1977). Astrophys.J.Lett., 217, L125.
Ostriker, J.P., Peebles, P.J.E. & Yahil, A. (1974). Astrophys.J. Lett., 193, L1.
Ostriker, J.P. & Tremaine, S. (1975). Astrophys.J.Lett., 202, L113.
Ostriker, J.P. & Turner, E.L. (1979). Astrophys.J., 234, 785.
Peebles, P.J.E. (1980). The Large Scale Structure of the Universe. Princeton University Press, Princeton, N.J.
Rood, H. & Leir, A. (1979). Astrophys.J.Lett., 231, L3.
Rood, H., Page, T., Kintner, E. & King, I. (1972). Astrophys.J., 175, 627.
Roos, N. & Norman, C.A. (1979). Astron.Astrophys., 76, 75.
Sastry, G.N. (1968). Publ.Astr.Soc.Pac., 80, 252.
Sastry, K.S. (1972). Astrophys.Sp.Sci., 16, 284.
Sastry, K.S. & Alladin, S.M. (1970). Astrophys.Sp.Sci., 7, 261.
Schechter, P. (1976). Astrophys.J., 203, 297.
Spitzer, L. (1958). Astrophys.J., 127, 17.
Toomre, A. (1977). In The Evolution of Galaxies and Stellar Populations, ed. B.M. Tinsley and R.B. Larson, p. 401. Yale University Press.
Toomre, A. & Toomre, J. (1972). Astrophys.J., 178, 623.
Tremaine, S. (1976). Astrophys.J., 203, 72.
Tremaine, S. & Richstone, D. (1977). Astrophys.J., 212, 311.
Villumsen, J.V. (1980). Mon.Not.R.Astr.Soc., submitted.
Visvanathan, N. & Sandage, A. (1977). Astrophys.J., 216, 214.
White, R.A. (1978). Astrophys.J., 226, 591.
White, S.D.M. (1976). Mon.Not.R.Astr.Soc., 174, 19.
White, S.D.M. (1978). Mon.Not.R.Astr.Soc., 184, 185.
White, S.D.M. (1979a). Mon.Not.R.Astr.Soc., 189, 831.
White, S.D.M. (1979b). Astrophys.J.Lett., 229, L9.
White, S.D.M. (1980). Preprint.
White, S.D.M. & Rees, M.J. (1978). Mon.Not.R.Astr.Soc., 183, 341.
White, S.D.M. & Valdes, F. (1980). Mon.Not.R.Astr.Soc., 190, 55.

THE STRUCTURE OF BARRED GALAXIES

John Kormendy

Dominion Astrophysical Observatory
Herzberg Institute of Astrophysics

1. INTRODUCTION

In this paper we review observational work on the stellar structure of barred galaxies. An underlying theme will be the growing realization in recent years that bars cause a variety of secular evolution effects in galaxy dynamics. Formerly, the long timescale of evolution through two-body encounters suggested that galaxies settle into a steady state after an initial collapse or merger phase (perhaps augmented by later infall), and thereafter evolve only their stellar and gaseous content. However, any collective phenomenon which produces a significant departure from an axisymmetric gravitational potential allows individual stars to interact with a coherent pattern. In this way a bar can interact with every other component in the galaxy. This can change both the dynamics and the stellar content on timescales intermediate between a collapse time and the age of the universe. Bars therefore take on a new importance in the study of galaxy evolution: since axial symmetry is turning out to be the exception rather than the rule (Freeman 1970; Kormendy 1980a; §2.1, below), secular evolution may have an important effect on the structure of most galaxies.

In section 2 we discuss the basic morphology of SB structure, and introduce the distinct components associated with bars (see Kormendy 1979b for a review of the component approach). Following sections discuss the dynamics and origin of these features. The conclusion lists the evolutionary effects identified to date. To keep this paper manageable, we restrict ourselves almost entirely to stellar-dynamical phenomena. Even so, this review is incomplete; a more thorough discussion will be given by Kormendy (1980c). Gas dynamics of barred galaxies have been reviewed by Roberts (1979a,b).

2. MORPHOLOGY OF BARRED GALAXIES

Figure 1 illustrates the basic components associated with bars

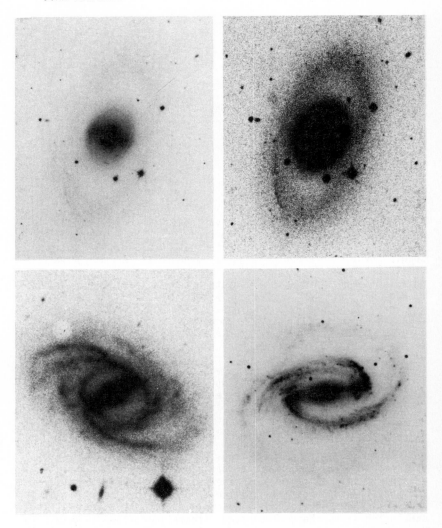

Fig. 1 - Examples of barred galaxies. (a, top left) The
(R)SB(lens)0+ galaxy NGC 3945. (b, top right) NGC 3945 printed at
higher contrast to show the outer ring. (c, bottom left) The
prototypical SB(r) galaxy NGC 2523. (d, bottom right) NGC 1300,
illustrating the alternative SB(s) structure. Emulsions and
exposure times are given in Kormendy (1979a).

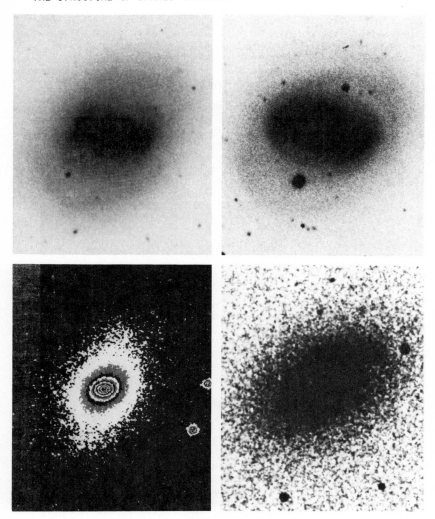

Fig. 1 (continued) - (e, top left) NGC 936 is an SB(rℓ)0 galaxy with a high-contrast bar and an unusually bright disk. (f, top right) NGC 4596 contains a lens broken at the end of the bar; cf. NGC 5101 in the Hubble Atlas. (g, bottom left) An isodensity tracing of NGC 2950, a SB(lens)0 galaxy with a triaxial bulge. The bulge is elongated almost horizontally (east-west), while the bar is shown by the outer contour. (h, bottom right) A high-contrast print from the Palomar Observatory Sky Survey shows that the disk position angle is between those of the bulge and bar.

(see also Sandage 1961, hereafter the Hubble Atlas; Kormendy 1979a). We emphasize that there is nothing unusual about any of the features shown. However, the galaxies display these features with abnormal strength and regularity. Such objects are ideal for the dynamical studies described in later sections.

2.1. Distinct Components in Barred Galaxies

Bars are distinguished from other triaxial features such as lenses and oval disks by their larger axial ratios and by the qualitative difference between their major- and minor-axis profiles (Fig. 1). Along the major axis the surface brightness is nearly constant interior to a sharp outer edge, while along the minor axis the profile is steep, as in a bulge. This behaviour distinguishes bars from elliptical galaxies, which are also believed to be triaxial ellipsoids. In fact, it is important to note that barred ellipticals do not exist - whenever a galaxy has a bar, it also has a disk. This association with high angular momentum material foreshadows the dynamical difference between bars and ellipticals found in §3: bars rotate much more rapidly than ellipticals.

Little photometry of bars is available. In published data, the fraction of light in the bar varies from 0% to $\sim$ 30% (de Vaucouleurs 1975; Crane 1975; Benedict 1976). This has suggested that bars constitute only a small perturbation on a basically axisymmetric, very centrally concentrated structure (King 1975). However, kinematic measurements discussed in §3 show that bars can greatly perturb the dynamics.

Triaxial bulges elongated perpendicular to the bar are seen in a number of galaxies (Fig. 1a,g,h), while other bulges are normal (Fig. 1e,f). These features will be discussed in §4.

Lens components (Hubble Atlas) are common features in early-type barred galaxies. A lens is an elliptical (as opposed to bar-shaped) feature occurring between the bulge and the disk, which has a shallow brightness gradient interior to a sharp outer edge, and then a much steeper outer gradient. We differentiate lenses (S0-a galaxies), which have rotation curves rising with radius, from the morphologically similar oval disks (Sb-m galaxies; e.g., NGC 4736, Kormendy 1979a), which generally have flat rotation curves (Kormendy and Norman 1980). Excellent examples of lenses are present in NGC 3945 (Fig. 1a,b) and in NGC 4596 (Fig. 1f). Note in Fig. 1b that the lens is only slightly larger in the deeper photograph, illustrating the sharpness of its outer edge. Other lenses are illustrated in the Hubble Atlas, Kormendy (1979a), and Sandage and Brucato (1979). Although lenses are especially common in barred galaxies, they also occur in some unbarred objects (e.g., NGC 1553, Freeman 1975;

THE STRUCTURE OF BARRED GALAXIES

Sandage and Brucato 1979).

Remark: The Revised Morphological Types of de Vaucouleurs, de Vaucouleurs and Corwin (1976, hereafter RC2, and references therein) do not distinguish lenses and inner rings (below), but denote both as (r). Furthermore, all SO galaxies are called "lenticulars" despite the fact that many SBOs and most SAOs do not have lenses. Thus one cannot determine from the RC2 whether or not a galaxy has a lens. Following Kormendy (1979a), we denote lenses by (ℓ).

Inner rings (r): NGC 2523 (Fig. 1c) contains a prototypical inner ring. Such rings are always the same size as the bar. As suggested by Hα photography (Hodge 1974) and color photography (Wray 1979), they have the same population as the disk. This is true even when the bar lacks young stars, as in NGC 2523. Sometimes inner rings occur at the rim of a lens (NGC 936, Fig. 1e); in later-type galaxies there usually is no lens (NGC 2523).

Outer rings (R) such as the one in NGC 3945 (Fig. 1b) are relatively dark inside, and are approximately tangent to the longest dimension of the B(lens) combination, i.e., the bar major axis. They are broader than inner rings, and are distinguished from them by their well-defined larger size, $< r_{(R)}/r_B > = 2.21 \pm 0.02$ ($\sigma/\sqrt{13}$), see Kormendy (1979a). Outer rings are generally largest in the direction perpendicular to the bar. A specific example is the face-on galaxy NGC 1291 (Mebold et al. 1979), whose outer ring has a true axial ratio of 0.90 ± 0.02. Some outer rings are more elongated than this (Kormendy 1979a), but others are certainly almost round, and it is not excluded that a small fraction may be elongated parallel to the bar.

Kormendy's (1979a) survey found true outer rings in only ~ 6 of 121 galaxies. However, there exist related features which are much more common. These include pseudo-outer rings (R'), which consist of spiral arms that almost close into a ring. Also common are outer disks which, like lenses, have shallow brightness gradients interior to a fairly sharp outer edge (e.g., NGC 4596, Fig. 1f). In all of these galaxies, the outer feature is similar in size and shape to true outer rings (Kormendy 1979a). That is, no part of an SB galaxy is generally circular. Furthermore, there exist many SA galaxies which are globally oval without any sign of a bar. (Bosma 1978; Kormendy 1979a; Kormendy and Norman 1979; see Kormendy 1980a for a review). Thus the fraction of all disk galaxies which are triaxial enough so that secular evolution might be important is rather larger than the 65% (de Vaucouleurs 1963) which are classified as SB or SAB.

This completes our introduction to the components found in barred galaxies. The stellar dynamics of these features are discussed in the following sections. Rings are made by gas-dynamical processes; they will be mentioned only briefly in §6.

2.2. Frequency of Incidence of Inner Rings and Lenses

The importance of inner rings and lenses to SB dynamics is illustrated by Table 1, which gives the percentages of galaxies of various Hubble types which contain these features (Kormendy 1979a). Table 1 shows that lenses are components of major importance in early-type barred spirals, while inner rings are similarly prominent in late-type systems. Specifically, over half of all SB0-a galaxies have lenses. Some of these also contain a ring; however, pure inner rings are not common in SB0-a galaxies. In contrast, no survey object of type SBab-c had a dominant lens. Instead, 63% had inner rings. If the dynamically similar pseudorings are included, fully three-quarters of late-type barred galaxies had such features. We conclude that the processes leading to lens and ring formation are not unusual, but are an integral part of SB dynamics.

Some galaxies of all types have neither a ring nor a lens. In early-type systems the disk is usually featureless. Late-type disks are dominated by a coherent, two-armed spiral pattern beginning at the ends of the bar (Fig. 1d). This SB(s) structure is especially common in SBcd-d galaxies. This may be due to their systematically lower masses (Kormendy 1979a), since the sophistication of galaxy structure generally correlates well with mass (e.g. luminosity classes of van den Bergh 1960).

Table 1. Frequency of Incidence of Inner Rings and Lenses

Feature	Type	0-a	ab-c	cd-d
r'		0	10	0
r		19	63	8
ℓ		54	0	0
Neither		27	27	92
Number of galaxies		37	49	13

2.3. Shapes of Lenses and Inner Rings

The three-dimensional shapes of lenses and inner rings have been discussed by Kormendy (1979a), Mebold et al. (1979), and de Vaucouleurs and Buta (1980). The best information comes from galaxies whose inclinations are known precisely. For example, in the SB(ℓ)0 galaxy NGC 1291, Mebold et al. (1979) measure a maximum HI rotation velocity of only 20 ± 5 km s^{-1}, implying that the inclination is 6° ± 2°. Since the galaxy is almost face-on, the true "equatorial" axial ratio of the lens is the observed value, 0.81 ± 0.02. Other galaxies with known inclinations are discussed by Kormendy (1979a). Also, kinematic detections of triaxiality have been reported in the lenses of NGC 1326 (Mebold et al. 1979) and NGC 4596 (Kormendy 1980b; see §3.2, below). Finally, we can study the frequency distribution of the apparent axial ratios of inner rings (Fig. 4 of Kormendy 1979a), since the common occurrence of inner rings around the rims of lenses shows that these features have the same shapes (de Vaucouleurs and Buta 1980). Combining all of these data suggests that lenses and inner rings have preferred "equatorial" axial ratios of $\sim$ 0.9 ± 0.05, with the bar filling the longest dimension (cf. Kormendy 1979a; de Vaucouleurs and Buta 1980). A few may be round. Most cannot be as elliptical as b/a = 0.80, or the implied axial ratio distribution would predict too few observed values with b/a > 0.85.

Detailed information on the thickness of lenses is available for only one galaxy, the edge-on S0 NGC 4762. As shown by Tsikoudi (1977) and Burstein (1979), the major-axis profile consists of three prominent plateaus in surface brightness. This is the signature of an edge-on SB(lens) structure. The conclusion of Tsikoudi and Burstein is that the lens of NGC 4762 is as thin as any SA0 disk. The same is true in NGC 4111 (Burstein 1979), although the lens morphology is more uncertain. The structure of edge-on galaxies is generally consistent with this conclusion. Apparently most lenses and bars are as flat as disks (see also Freeman 1979). However, more detailed data on the structure perpendicular to the plane would be valuable.

3. BARS

Our discussion of bar dynamics is weighted toward observational results. N-body work is reviewed briefly. Theoretical dynamics of bars are introduced only via plausibility arguments; more detailed discussions can be found in Lynden-Bell (1979), Contopoulos (1979, and references therein), and Kormendy (1980c).

3.1. Basic Properties of Bars: Theoretical Results

The basic nature of bars is well described by the analogy of kinematic density waves (Kalnajs 1973; Toomre 1977). This requires a digression on the principal resonances of individual stars with global patterns.

3.1.1. The Principal Resonances. The galactic orbit of a star in nearly circular motion is generally an unclosed rosette (Fig. 1a of Kalnajs (1973). That is, there are a non-integral number of radial oscillations at "epicyclic" frequency κ for every rotation at angular velocity Ω about the center. However, there are a large number of rotating frames of reference in which the orbit appears closed. The simplest of these rotates about the nucleus at $\Omega_p = \Omega$ ("corotation", hereafter CR). In it the star describes a small epicyclic ellipse centered on its mean radius. The next-simplest such rotating frames have $\Omega_p = \Omega - \kappa/2$ ("inner Lindblad resonance", ILR) and $\Omega_p = \Omega + \kappa/2$ ("outer Lindblad resonance", OLR). In the ILR frame the orbit is approximately elliptical and centered on the nucleus. That is, there are two epicyclic oscillations, and therefore two apogalactica, for every stellar revolution. The importance of ILR stems from this bisymmetry. If a bisymmetric pattern such as a bar rotates at $\Omega_p = \Omega - \kappa/2$, then every time a star is at a certain phase of its radial oscillation, it is at the same position with respect to the pattern. For example, if the star is at apogalacticon when it is on the crest of the bar, then, as it drifts ahead of the pattern (because $\Omega > \Omega_p$), the next time it is at apogalacticon it is again on the crest of the bar (but on the other arm). Similarly, a star at OLR drifts backwards to the other arm during one radial oscillation. In either case the bar repeatedly perturbs the orbit in the same way, producing large secular changes. Thus, at the three principal resonances important interactions can take place between individual stars and the global pattern. As a result, any secular evolution, whether stellar or gas-dynamic, tends to proceed fastest at the resonances (e.g., see §5).

3.1.2. <u>Bars as Almost-Kinematic Density Waves</u>. The well-known solution to the structural shearing produced by differential rotation is to make the pattern out of a density wave which rotates at constant Ω_p while stars drift through it at Ω. A crucial circumstance favouring the production of bisymmetric waves was discovered by Lindblad (1956) and exploited more fully by Kalnajs (1973) and Toomre (1977). Lindblad noted that $\Omega - \kappa/2$ is small and nearly constant over a considerable range of radii in typical galaxies. As emphasized by the above authors, if $\Omega - \kappa/2$ were exactly constant, then a bar could be constructed out of ILR orbits by aligning all of their major axes (Fig. 3a of Kalnajs 1973). The ellipses would precess together and maintain the shape of the bar. In reality, $\Omega - \kappa/2$ is not exactly constant, so even the best choice of Ω_p would result in a slow shearing. Therefore the remaining requirement on the bar-shaped potential is that it provide the extra persuasion needed to make the orbits precess not at $\Omega - \kappa/2$ but at some exactly constant rate $\Omega_p \sim \Omega - \kappa/2$ (Toomre 1977).

The above task is easiest in a galaxy whose rotation curve rises slowly to a maximum value near the edge of the disk, because then $\Omega - \kappa/2$ is small and slowly-varying (e.g., M33, Fig. 4a of Shu, Stachnik and Yost 1971). On the other hand, if there is a large bulge, as in M81, then $\Omega - \kappa/2$ is large and rapidly varying over a substantial part of the disk (Fig. 4b,c of Shu et al. 1971). This makes plausible the conclusion below that it is much easier to maintain a bar (or spiral structure - see Kormendy and Norman 1979) in a galaxy with little central mass concentration.

3.1.3. The Pattern Speed. The pattern speed of a bar is unknown in principle and not directly measurable in practice. We have seen that it is advantageous to have $\Omega_p \sim \Omega - \kappa/2$. This section summarizes theoretical and experimental results which suggest that Ω_p is close to but slightly larger than the maximum value of $\Omega - \kappa/2$ in the bar, so that any ILR is confined to the bulge.

The orbits which make up a bar can be classified into families each characterized by a progenitor orbit which is strictly periodic (Contopoulos 1979). It is therefore sufficient to study the orientations of the periodic orbits in a bar field, and to ask whether the response to an incipient bar will reinforce it and help it to grow. If the departures from circular motion are small, the periodic orbits are readily calculated (Barbanis and Woltjer 1967; Sanders 1977; Contopoulos and Mertzanides 1977, and references therein). Our discussion follows Sanders (1977). In the bar frame, the departure from circular motion is

$$\delta r = c_1 + c_2 \cos [2(\Omega - \Omega_p)t], \qquad (1)$$

where t is time, c_1 is a constant, and where

$$c_2 = [p + q\Omega/(\Omega - \Omega_p)] [\kappa^2 - 4(\Omega - \Omega_p)^2]^{-1} \qquad (2)$$

(Sanders and Huntley 1976; Sanders 1977). In the above, p and q are the radial and tangential amplitudes of the nonaxisymmetric force. The orbit (1) is an ellipse with the orientation of its major axis determined by the sign of the amplitude c_2. Equation (2) shows that the orientations of the periodic orbits in a bar potential change abruptly by 90° at each of the principal resonances ILR, CR and OLR. If the galaxy is very centrally concentrated, a single effective ILR is unavoidable. Interior to this the response is perpendicular to the bar and so opposes it. On the other hand, if $\Omega_p > \Omega - \kappa/2$ everywhere, so that there is no ILR, then all of the orbits interior to corotation will align with the bar and help it to grow. Furthermore, the elongation of the periodic orbits is largest close to a resonance (see equation 2). Thus the most favourable pattern speed is one which almost has an ILR, so that the bar-like response has high amplitude.

A variety of numerical experiments follow the above plausibility argument in predicting that $\Omega_p > (\Omega - \kappa/2)_{max}$. These include the n-body models of Sanders (1977), Zang and Hohl (1978), Miller and Smith (1979a,b; see Huntley 1980a,b), and Sellwood (1980a,b). Similarly, calculations by Bardeen (1975) of global

instabilities in gas disks show that the modes just fail to have
an ILR. In a different sort of calculation, Sanders and Tubbs
(1980) investigate the range of parameters which make gas respond
realistically to an imposed bar. Their preferred value of $\Omega - \kappa/2$
is just slightly smaller than the maximum $\Omega - \kappa/2$. Finally,
Lynden-Bell's (1979) particle-orbit description of a bar
instability also requires $\Omega_p \sim \Omega - \kappa/2$. These studies leave us
with strong hints but no firm conclusion. Pattern speeds which
just fail to allow an ILR are preferred. However, some models
have Ω_p several times as large as $(\Omega - \kappa/2)_{max}$ (Sellwood
1980a,b). Also, at least one model actually has an ILR (Zang and
Hohl 1978), and some of the kinematic observations discussed in
§3.2 also imply that there is an ILR. However, in all of these
cases the disk outside the bulge has little enough central mass
concentration so $\Omega - \kappa/2$ is small. We therefore adopt the
provisional assumption that Ω_p is slightly greater than the
maximum $\Omega - \kappa/2$ over the region of the bar. In applying this
assumption, it is important to remember that the velocity
dispersion is usually large (§3.2), so $V(r)$ cannot directly be
used to derive $\Omega - \kappa/2$ (as in Kormendy and Norman 1979). However,
Huntley (1980a,b) has shown that the azimuthally averaged gaseous
velocity field does properly measure the potential, at least in
Miller and Smith's (1979a,b) bars.

3.1.4. <u>The Length of the Bar</u>. The characteristic, sharp ends of
bars are potentially powerful diagnostic tools for deciphering SB
dynamics. We therefore need to determine why bars end so
suddenly, and especially, to ask where bars end with respect to
the resonances. A variety of hints suggest that the end of the
bar is determined by the rotation curve (i.e., the mass
distribution). Observed, theoretical and numerical bars tend to
end where the rotation curve levels off to constant velocity, and
in any case do not extend beyond corotation.

The possible extent of bars has been investigated
theoretically by Lynden-Bell (1979) and by Contopoulos and
collaborators (Contopoulos and Mertzanides 1977; Contopoulos 1979;
Contopoulos and Papayannopoulos 1980, and references therein).
The program of Contopoulos aims to derive bar models in three
stages: (i) find the form of the orbits in a bar potential; (ii)
calculate the density enhancements produced by the orbits, and
(iii) construct self-consistent models by demanding that the
response density equal the imposed bar. The strongest conclusion
to emerge is that bars do not extend past corotation. Instead,
many orbits beyond corotation are unstable if the bar is strong;
i.e., stars gain angular momentum from the bar and spiral outward.
A bar can exist everywhere interior to corotation except between
two inner Lindblad resonances. Bar formation is therefore easiest
if there is no ILR: the bar then extends all the way from the
center to corotation. In contrast to this, Lynden-Bell (1979)
suggests that the bar-forming region is more restricted. He

94

investigates the conditions which allow an incipient bar rotating at Ω_p to capture additional orbits precessing at $\Omega - \kappa/2 \sim \Omega_p$. The required condition turns out to be that $\Omega - \kappa/2$ increases with increasing orbital angular momentum. This occurs approximately over the rising part of the rotation curve, and especially between the two inner Lindblad resonances. The above approaches differ in their basic assumptions. Lynden-Bell uses very elongated orbits while Contopoulos assumes that the motions are almost circular. Secondly, Lynden-Bell assumes that bar orbits never leave the bar, while the orbits in Contopoulos' picture can be uncaptured (see Lynden-Bell 1979). It is important to determine which picture is closer to the truth, since the fundamental nature of the bar phenomenon is at issue.

Kormendy and Norman (1980) have made a preliminary attack on this problem as part of a comparison of bar-formation theories with observations. They find that observed rotation velocities are generally rising out to the end of the bar, and constant in the rest of the disk (e.g., NGC 7723, Chevalier and Furenlid 1978; an exception is NGC 5383, Peterson et al. 1978; Sancisi, Allen and Sullivan 1979). In agreement with this result, calculations of global instabilities in gas disks show that the modes are bar-shaped only over the rising part of the rotation curve (Bardeen 1975; Iye 1978; Aoki, Noguchi and Iye 1979). At the velocity turnover radius, the modes turn abruptly from bar-shaped to spiral (see especially Fig. 3 and 4 of Bardeen 1975). The ends of these bars are somewhat interior to corotation. All this agrees well with Lynden-Bell (1979). However, Sanders and Tubbs (1980) find that the gas response to an imposed bar is most realistic when the bar ends near corotation. Also, in n-body models the bar extends precisely to corotation (e.g., Miller and Smith 1979b; Hohl and Zang 1979, and especially Sellwood 1980a,b). However, in all of these models corotation occurs just beyond the velocity turnover radius. Thus the predictions of Contopoulos and Lynden-Bell may in practice be less inconsistent than they seem. In particular, bars may well be made up of both of the two kinds of orbits discussed above. In the rest of this paper we follow the numerical results and assume that bars end at corotation.

This completes our theoretical discussion of the nature of bars, and of their fundamental parameters. A more extensive review can be found in Kormendy (1980c).

3.2. Stellar Velocity Fields

Observational progress on deciphering bar dynamics requires the measurement of stellar velocity fields. Little such data are currently available. Kormendy (1980b) has therefore begun a program to map velocities and velocity dispersions in early-type barred and related galaxies. The aims are to search for non-circular streaming motions, and to examine the balance between rotational support and heat. Comparisons with numerical models suggest which theoretical approaches correctly incorporate the physics of bar formation. Comparisons with the kinematics of other components put constraints on any secular evolution processes. The galaxies are chosen to be prototypical, high-luminosity systems. The sample is weighted toward SB0 objects to avoid complications from gas dynamics, and to ensure that the galaxies are as transparent as possible. Measurements were obtained with the High Gain Video Spectrometer (Kormendy and Illingworth 1980) and the Kitt Peak National Observatory 2.1m and 4m telescopes. The limiting surface brightness is $\sim$ 24 B mag arcsec^{-2} in a 2-3 h integration. Velocities and dispersions were calculated using a Fourier quotient program written and kindly made available by P. Schechter. The reduction process is described by Schechter and Gunn (1979), Fried and Illingworth (1980), and Kormendy and Illingworth (1980). For convenience, we refer to the resulting run of velocity along a spectrograph slit as a rotation curve $V(r)$, recognizing that V does not directly measure mass when the dispersion σ is large. Interpretations are based on the assumption that all components except the bulge are flat. Then the local ratio of V/σ is independent of inclination for bars, lenses and disks.

NGC 936 (Fig. 1e) contains the strongest bar which Kormendy (1980b) has studied to date. In contrast to many SB0s, which have virtually no outer disk, the surface brightness outside the B(rℓ) structure is high enough for reliable measurements. For these reasons, NGC 936 is an ideal object for dynamical studies. Fig. 2 shows rotation curves and velocity dispersions along the major axis and the bar. These data provide the first clear detection of non-circular stellar motions in a bar.

(i) Outside of the rapidly rotating bulge, the major-axis rotation curve rises linearly to the edge of the combined ring and lens, at r = 51". Here there is a jump in velocity of 50 - 70 km s^{-1}. Tests of the data suggest that the discontinuity is real. If so, it is evidence for a sudden change in the shapes of the orbits at the edge of the lens.

(ii) Along the bar the velocity center is displaced by Δr = 4".8 from the optical center. Such a shift dramatically reduces the scatter of the points, revealing a complicated velocity structure which is precisely reproduced on both sides of the galaxy. Note that the bulge does not require such a shift: it rotates about its center. The bulge is not offset from the photometric center of the bar. The simplest interpretation is that the bar's velocity field consists of slightly asymmetric, very elongated streamlines. If the orbits are elongated, as they are in the gas velocity field of NGC 5383 (Fig. 6b of Peterson et al. 1978), then even a slight asymmetry combined with a slit centering error of < 1" will produce a large apparent shift of the

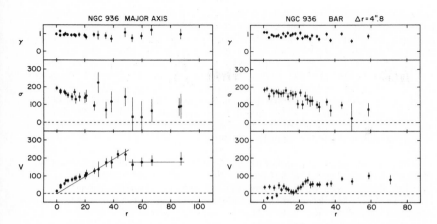

Fig. 2 - Velocity V, dispersion σ and linestrength γ along the major axis (left) and bar of NGC 936. Velocity is in km s^{-1} and radius in arcsec. The linestrength is an approximate ratio of the overall strength of lines in the galaxy and in the KO-2 III standard star used. Since different standard stars are used for different observations in this paper, comparisons of γ in different panels are generally not possible. The horizontal line represents the weighted mean velocity of the outer four major-axis points. The other line is an unweighted least-squares fit to the outer six points interior to the velocity discontinuity and constrained to pass through the origin. The radius of the lens at left is 51"; the length of the bar at right is 42".

kinematic center along the slit. Thus the above effect is evidence for systematic non-circular streaming motions in the bar.

(iii) No value of the inclination (plausible or otherwise) will transform the major-axis rotation curve into that along the bar. Note especially that the rotation curve along the bar is far from rigid, while that along the major axis is linear. This again is evidence for non-circular streaming. We suspect that the very differential rotation observed along the bar of NGC 1300 by Peterson and Huntley (1980) is due to a similar effect.

(iv) Velocity dispersions comparable to the rotation velocity are seen in both the bar and the lens.
These observations imply that the strong bar of NGC 936 has a large effect on the dynamics, producing a complicated, non-circular velocity field.

It is interesting to compare the amount of random motion in numerical bars with the above observations. For example, Hohl and

Zang (1979) tabulate the ratio V_{max}/σ_0 of the maximum projected rotation velocity to the projected central dispersion for various orientations of their bar model IV. The geometry most similar to NGC 936 has $V_{max}/\sigma_0 = 0.49$. The observed value along the bar is ~ 0.41. Locally, V/σ is typically < 0.3. Comparisons with the Miller and Smith (1979b) model show similar consistency. Thus the degree to which rotation and random motions support the bar is similar in the observations and in the models. That is, these models, which do not (as originally intended) describe elliptical galaxies because they rotate too rapidly (§4), turn out to be reasonable first-order models of bars.

To investigate the generality of the above effects, Kormendy (1980b) has studied the SB(lens)0 galaxy NGC 4596, which is similar to NGC 936 both in morphology and in viewing geometry (Fig. 1e,f). The kinematics of NGC 4596 turn out to closely resemble those of NGC 936. The kinematic center of the bar is again displaced from the nucleus, this time by only 2". The major-axis rotation curve again projects poorly onto the bar. The velocity dispersion in the bar is large: $V_{max}/\sigma_0 = 0.58 \pm 0.06$ compared to $V_{max}/\sigma_0 \sim 0.50$ in Hohl and Zang's (1979) model IV. Interestingly, rotation of up to 50 km s^{-1} is seen along the minor axis, showing that the disk is oval. This is consistent with the fact that the lens and outer disk have different ellipticities and different major-axis position angles (Fig. 1f).

Finally, NGC 3945 is a type example of an (R)SB(lens)0 galaxy, illustrated in Fig. 1a,b. Fig. 3 shows its major- and minor-axis rotation curves. Since the bar is oriented along the minor axis, and since the major axes of oval features in SB galaxies are either parallel or perpendicular to the bar, there should be no difference between the photometric and kinematic minor axes. In confirmation of this picture, the minor axis shows no rotation. The radial velocity dispersion at $r \sim 23"$ along the bar is 135 ± 13 km s^{-1}, not much smaller than the maximum lens rotation of 224 ± 13 km s^{-1}. Since the radial and azimuthal dispersions are comparable, the stellar orbits are again far from circular.

We conclude that well-developed bars constitute a major perturbation on the stellar dynamics. There is evidence both for large random motions and for systematic non-circular streaming. In terms of the general balance between rotation and heat, the observed bars are similar to numerical models (e.g., Miller and Smith 1979b; Hohl and Zang 1979; Sellwood 1980a,b). These are therefore useful as preliminary bar models.

4. TRIAXIAL BULGE COMPONENTS

It has been known for some time that the bulges of a few SB galaxies are elongated into secondary nuclear bars (NGC 1291 and

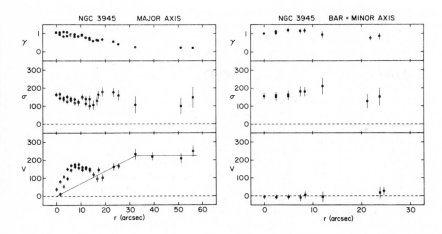

Fig. 3 - Rotation and velocity dispersion along the major and minor axes of NGC 3945. The straight lines are calculated as in Fig. 2. The radius of the lens is 51", and of the bar, 34". The asymmetry of the rotation curve near the center of the major axis is due to the repulsion of the ISIT readout electron beam by the charge distribution; this effect is readily correctable (Kormendy 1980b).

1329, de Vaucouleurs 1974, 1975). Additional examples have been found by Kormendy (1979a), Sandage and Brucato (1979) and Schweizer (1980). Kormendy and Koo (1980) find that these features are very common in a large-scale photographic survey of the 35 brightest northern SB0--SBab galaxies. The two bars are generally perpendicular (Kormendy 1979a; Schweizer 1980; Kormendy and Koo 1980). Triaxial bulges are of interest because they illuminate the nature of bar dynamics, and because they are another triaxial component which may be involved in secular evolution.

Fig. 1 g,h illustrate a particularly well-developed triaxial bulge, NGC 2950 (Kormendy and Koo 1980). The position angles of the bar, bulge and disk are all different. Especially important is the fact that the disk position angle is between those of the bar and the bulge. This means that the bulge position angle is not different from that of the disk only because it is contaminated by bar light. Rather, the bulge must be prolate. Since the shortest dimension in these objects is probably the one perpendicular to the disk (Kormendy and Koo 1980), we refer to these bulges as triaxial.

Another excellent example of a triaxial bulge is seen in

JOHN KORMENDY

NGC 3945 (Fig. 1a; see also Fig. 2c,d of Kormendy 1979a). Conveniently, the bar is oriented along the minor axis, so the bulge is elongated along the major axis. It is therefore well measured by the rotation curve of Fig. 3. This bulge rotates very rapidly. In fact, $V_{max}/\sigma_0 = 1.03 \pm 0.08$; $V_{max}/\langle\sigma\rangle = 1.22 \pm 0.05$, where $\langle\sigma\rangle$ is the luminosity-weighted mean dispersion. Since the bulge is only E3.3 in shape, it rotates faster in terms of σ than do oblate-spheroid dynamical models which owe their shape to rotation (Binney 1980, and references therein). This result, and rotation data on three more triaxial bulges, are shown in Fig. 4. For comparison, data are also shown on normal bulges (Kormendy and Illingworth 1980; Kormendy 1980b; Illingworth 1980) and ellipticals (Illingworth 1980, and references therein). The diagram is presented in terms of both the central and mean dispersions. For ellipticals, only V/σ_0 values are published; these are shown in both panels. The same is true of the Hohl and Zang (1979) bar model. However, comparison with the oblate and prolate model lines is more appropriately made using the mean dispersion (Binney 1980).

Fig. 4 shows that ordinary bulges rotate faster than typical ellipticals (Kormendy and Illingworth 1980). They generally satisfy the relation predicted for oblate-spheroid models. Two ordinary bulges in the barred galaxies NGC 936 and NGC 4596 behave like their SA counterparts. However, the triaxial bulges rotate significantly faster than the ordinary ones, and marginally faster than the oblate models. The effect is large in NGC 3945, as noted above. In rotating rapidly, triaxial bulges resemble the bar models illustrated. As noted by Miller and Smith (1979b), and Hohl and Zang (1979), there is a fairly wide range of viewing angles from which these models also rotate as fast as or faster than the oblate models.

A further similarity to bars is seen in the velocity anisotropy. While the ratio of radial to azimuthal dispersion is close to 1 in the ordinary bulge of NGC 4596, it is ~ 1.2 in the Hohl and Zang (1979) model, and 1.4 - 1.7 in the outer half of the bulge of NGC 3945 (Kormendy 1980b).

These observations imply the following conclusions (Kormendy 1980b).

(i) Kinematic observations and n-body models show that bars are dynamically different from elliptical galaxies, which are also commonly believed to be triaxial. Bars are built up of much higher angular momentum orbits (Binney 1979). This is perhaps not surprising in view of the result of §2 that bars are seen only in the presence of a disk.

(ii) Triaxial bulges in SB galaxies resemble bars rather than ellipticals.

100

(iii) This suggests that ellipticals, and ordinary and
triaxial bulges form a sequence of increasing rotation.
Kinematics and photometry of ordinary bulges suggest that they are

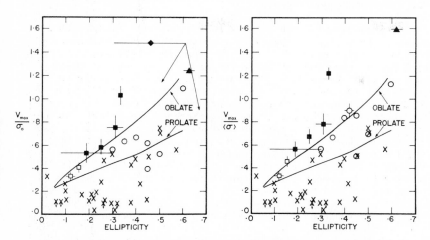

Fig. 4 - Ratio of the maximum rotation velocity V_{max} of the
ellipsoidal component to the central (left) and mean velocity
dispersions. The symbols used are as follows: X, elliptical
galaxies; circles, bulges of SA galaxies; squares, bulges of SB
galaxies; triangles, bar model of Miller and Smith (1979b);
diamond, model IV of Hohl and Zang (1979). Sources of galaxy data
are given in the text. Closed symbols refer to triaxial bulges or
models, open symbols to ordinary bulges. The V/σ error bars
(shown only for bulges taken from Kormendy 1980b) are estimates
formed by combining the formal errors in V and σ given by the
Fourier quotient program. Horizontal lines are not error bars,
but represent the range of ellipticity observed. When this is
large (NGC 2859; Hohl and Zang model) the maximum ellipticity
occurs at a very different radius from the maximum velocity
(ε at V_{max} is the left-hand end of the line). Both bar models are
shown as seen from the equatorial plane and with the bar
perpendicular to the line of sight. For the Hohl and Zang model,
we indicate how V/σ_0 and ε vary when the bar is viewed at 45° but
still from the equatorial plane (left-hand arrow) and at 45° to
the equatorial plane with the bar oriented either across the line
of sight or at 45° to the line of nodes. These arrows emanate
from the right-hand end of the ε "error bar" because only maximum
values of ε are quoted by Hohl and Zang (1979). Finally, the two
diagonal curves represent oblate-spheroid models which owe their
shape to rotation, and a class of prolate models rotating
end-over-end (Binney 1980). Changing the observed inclination
moves models parallel to the oblate line. The prolate line is a
median line for models distributed at all possible orientations.

spheroids. Apparently the extremes of the above sequence are
triaxial, but for different reasons in the case of bars (rapid
rotation) and ellipticals (initial velocity anisotropies, Binney
1980, and references therein).

The process which makes these bulges triaxial is not known.
However, the following questions suggest that these features may
provide a useful diagnostic tool for probing bar dynamics. First,
which bar is the "engine" for creating the very regular structure
which we see? Kormendy (1979a) argues that the normal bar is
responsible for evolution. However, the triaxial bulge of
NGC 3945 contains much more mass than the bar (although its mass
distribution is much less triaxial). Second, are triaxial bulges
always perpendicular to the bar? If not, if their pattern speeds
differ from those of bars, then ferocious dynamical friction
should result. This would be damaging to the health of both
components. If the two bars are always perpendicular, it is
tempting to interpret them in the context of the orientations of
closed orbits in a bar potential. Since these rotate by 90° at
ILR, there may be an ILR between the bulge and the bar. However,
exploration of the possible locations of the resonances shows that
it is difficult to interpret all of the features observed within
the above theoretical framework. Thus the existence of triaxial
bulges may lead to refinements in our picture of bar dynamics.

5. LENSES

We saw in §2 that 54% of SB0-a galaxies contain lenses.
Kormendy (1979a) has found that in virtually every case the bar
exactly fills the lens in one dimension. In objects with known
inclination, it is clear that the bar fills the longest dimension
of the lens. These and other morphological observations led
Kormendy (1979a) to suggest that lenses originate through a (still
unknown) process which makes some bars evolve away to a nearly
axisymmetric state. In this section we show that recent work on
SB(ℓ) dynamics supports the above hypothesis.

Because half of all early-type barred galaxies are transition
cases, any evolution process is probably secular. The assumption
is that an interaction with some other component gradually allows
stars to escape from the bar. Since lenses occur preferentially
in S0-a galaxies, this other component may be the bulge.

The interaction of bars and bulges has been studied in another
context by Sanders and van Albada (1979) and by Sellwood (1980a).
Sanders and van Albada show that a bar heats those bulge stars
which are in ILR with it. The heating is anisotropic, being small
perpendicular to the disk. Sanders and van Albada suggest that
the bar evolves away, leaving behind an anistropic velocity
distribution with slow rotation; i.e., an elliptical galaxy.

However, this interpretation seems unlikely, first because bars occur only in the presence of a disk, and second because SB bulges rotate very rapidly. Even if the bar is destroyed, the large amount of angular momentum in all components cannot be eliminated. In particular, the disk survives. Thus it is difficult to understand how bulge heating by a bar can make ellipticals. Rather, the process might very plausibly produce lenses.

To show how bar orbits are affected requires an examination of the transfer of angular momentum, since this is the quantity which is directly related to the shapes of the orbits. When the bulge is hot and nonrotating, bars are found to lose angular momentum to the bulge (Sellwood 1980a). Essentially, this is a dynamical friction produced by the fact that the pattern speed of the bar is larger than the precession rate $\Omega - \kappa/2$ of bulge orbits. The effect of decreasing the angular momentum of bar orbits is to make them more elongated (Lynden-Bell 1979). Thus a bar which interacts with a non-rotating bulge will grow stronger.

However, we saw in §4 that SB bulges rotate very rapidly (e.g., Fig. 3). Very probably they rotate rapidly enough so that most bulge stars precess faster than the bar. In other words, there is an ILR in the outer part of the bulge. Bulge stars which precess faster than Ω_p will feed angular momentum into the bar. These are the stars just inside ILR. Because of the strong density gradient, there are always more of these stars than stars just outside ILR, which form an angular momentum sink. Thus the bar gains angular momentum. This makes its orbits more circular (Lynden-Bell 1979). As the bar grows weaker, it becomes less capable of persuading all of its orbits to precess together at Ω_p. Some stars may therefore escape from the bar. The escaped stars and the fattening of the bar may combine to produce a nearly circular component with a radial brightness gradient similar to that of a bar, i.e., a lens.

This qualitative scenario has many attractive features which are in accord with the morphological evidence (Kormendy 1980c). For example, it predicts that lenses occur in galaxies with strong central concentrations and rapidly rotating bulges, as observed. Clearly a great deal of theoretical work is required to determine whether the suggestion is quantitatively tenable. Meanwhile, some observational tests are already possible. One prediction of almost any bar-lens evolution process is that lenses should be fairly hot. That is, we can ask whether lenses are "sufficiently similar" to bars.

The velocity data of §3 show that lenses in barred galaxies are in fact hot. A particularly good example is NGC 3945 (Fig. 3). The velocity dispersion of the bulge decreases with radius out to 18", where the bulge fades into the lens. Here the azimuthal dispersion increases, while the rotation velocity drops.

103

In terms of V/σ, the inner lens is considerably hotter than the bulge. At larger radii, the azimuthal dispersion decreases to small values at the edge of the lens. This is consistent with the bar-lens evolution picture.

A more conclusive test is provided by lenses in unbarred galaxies. Preliminary data for two such objects are shown in Fig. 5 (Kormendy 1980b). NGC 1553 is the best-developed and nearest SA(lens) galaxy known. Clearly its velocity dispersion behaves the same way as that of NGC 3945. In particular, at 15" radius, where the lens takes over the brightness profile (Freeman 1975), the velocity drops and the dispersion rises. The inner part of the lens of NGC 2784 is similarly hot.

The relative importance of rotation and heat is illustrated in Fig. 6 for all lenses studied to date. Also shown are two n-body models found in §3 to be good first-order models of bars. This figure leads to the following conclusions (Kormendy 1980b).

(i) Lenses in barred galaxies are as hot as models of bars. Both show similar, roughly linear increases of V/σ_θ with radius. At about half the radius of the lens, $V/\sigma_\theta \sim 0.8$. Since the radial dispersion σ_r is observed to be even larger than σ_θ, and since the ratio of rotational to random energy is proportional to $V^2/(\sigma_r^2+\sigma_\theta^2)$, we see that the inner parts of these lenses are largely supported by heat.

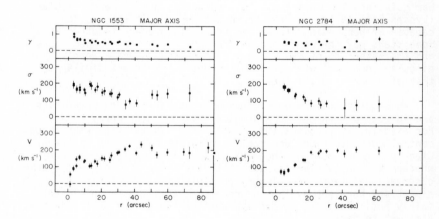

Fig. 5 - Major-axis velocity data for the SA(lens)0 galaxies NGC 1553 and NGC 2784. The radius of the lens is 36" at left and 40" at right. These data were obtained with the (photographic) image-tube spectrograph of the CTIO 4m telescope. Since the bulge of NGC 2784 is saturated on these spectra, its rotation curve does not extend to the center.

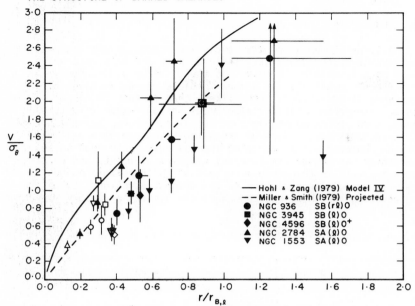

Fig. 6 - The relative importance of rotation and heat as a
function of radius in lenses of barred and unbarred galaxies.
V/σ_θ is the local ratio of rotation velocity to azimuthal velocity
dispersion. V/σ_θ error bars are determined by appropriately
combining V and σ error estimates shown in the major-axis rotation
curves of previous figures. Radius "error bars" show the range in
radii over which data were averaged into one point, which is
plotted at the mean radius of individual measurements weighted
inversely by the square of the velocity or dispersion errors.
Radii are normalized to the radius of the lens, r_ℓ (galaxies) or to
the corotation radius, r_B (models). Open symbols imply contamination by bulge light.

(ii) Lenses in two unbarred galaxies are kinematically
indistinguishable from SB lenses. That is, they are also hot.

(iii) The dispersion at the rim of a lens is small (see also
Bosma and Freeman 1980). This is expected, since if lenses are to
have sharp edges the radial dispersion must be small, and if bars
are to be narrow, the azimuthal dispersion at the end of the bar
must also be small.

It is worth emphasizing that these results are not due to
contamination of the lens dispersion by bulge light. Fig. 6 shows
that V/σ_θ is decreasing over the outer half of the bulge, and only
begins to rise again in the lens. Therefore, Fig. 6 provides
considerable support for the bar-lens evolution picture.

6. CONCLUSION

A large number of secular evolution processes are apparently at work in barred galaxies. Table 2 lists the main ones which have been identified to date. Of these, we have discussed only the second in any detail; a more complete review is given by Kormendy (1980c). It is very likely that further processes remain to be discovered. Table 2 also omits certain other effects of bars, such as the maintenance of spiral structure, and the nature of triaxial bulges. None of these processes are very well understood. However, it is becoming increasingly clear that these processes can change the internal dynamics of barred and oval galaxies. It is therefore important to further explore the effects of secular evolution on the present structure and on the observations which we use to deduce the history of galaxy formation.

Table 2. Secular Evolution in Barred Galaxies

	Process	Raw Material	End Product
1.	Heating by the bar (Hohl 1971)	Disk stars	Exponential disk
2.	Evolution of bar to lens (?)	Bar stars	Lens
3.	Dissipative rearranging of disk material (see Roberts 1979a,b)	Disk gas	Bulge and nucleus (4,5); inner ring; outer ring
4.	Building up of the bulge with disk material (?) (Kormendy 1980c)	Disk gas	Bulge
5.	Feeding Seyfert nuclei (?) (Simkin, Su and Schwarz 1980; Kormendy 1980c)	Disk gas	Nuclear activity
⋮	⋮	⋮	⋮

Stellar-dynamical studies have shown considerable progress toward an understanding of the bar phenomenon. We now have good first guesses of the basic parameters Ω_p and the length of the bar. It would now be helpful to have more detailed numerical models. Effects which these models do not yet show include the sharpness of the end of the bar, the flatness of the bar (i.e., dissipation during bar formation), and the production of several components, especially bulges with strong central concentrations. We emphasize that there is a particular need for the publication of more quantities which are directly comparable with observations. These include the run of density and of all components of velocity and velocity dispersion, both unprojected and projected at various viewing angles. Observationally, we need much more detailed velocity fields. It would be especially helpful to have stellar and gaseous velocity fields in the same objects, since the former probe the dynamics of bars, while the latter more directly measure the mass distribution. Finally, we need two-dimensional surface photometry as input for dynamical models. These programs are currently in progress.

Our understanding of bar dynamics is still far from complete. However, we now have the encouraging feeling that the subject is making rapid progress. In particular, we now have a good picture of what further observations and theory are needed to develop our understanding of the dynamics of barred galaxies.

ACKNOWLEDGEMENTS

The kinematic observations discussed here were obtained as a Visiting Astronomer at Kitt Peak National Observatory and at Cerro Tololo Interamerican Observatory, which are operated by the Association of Universities for Research in Astronomy, Inc., under contract with the National Science Foundation. It is a pleasure to thank P. Schechter for making available the Fourier quotient program, and G. Illingworth for technical support. Helpful discussions with J. Binney, J. Huntley, G. Illingworth, D. Lynden-Bell, C. Norman and J. Sellwood are gratefully acknowledged. M. Fall and D. Lynden-Bell exerted gentle but persistent pressure to cut the manuscript down to size. Finally, I thank M. Rees for his hospitality at the Institute of Astronomy, Cambridge, where much of this paper was written.

REFERENCES

Aoki, S., Noguchi, M., and Iye, M. 1979, Pub. A.S. Japan, 31, 737.

Barbanis, B., and Woltjer, L. 1967, Ap. J., 150, 461.

Bardeen, J.M. 1975, in IAU Symposium No. 69, Dynamics of Stellar Systems, ed. A. Hayli (Boston: Reidel), p. 297.

Benedict, G.F. 1976, A.J., 81, 799.

Binney, J. 1979, in Photometry, Kinematics and Dynamics of Galaxies, ed. D.S. Evans (Austin: Dept. Astron. Univ. Texas at Austin), p. 357.

Binney, J. 1980, in Structure and Evolution of Normal Galaxies, ed. S.M. Fall and D. Lynden-Bell (Cambridge, U.K.: Cambridge Univ. Press), in press.

Bosma, A. 1978, Ph.D. Thesis, Rijksuniversiteit te Groningen.

Bosma, A., and Freeman, K.C. 1980, in preparation.

Burstein, D. 1979, Ap. J., 234, 829.

Chevalier, R.A., and Furenlid, I. 1978, Ap. J., 225, 67.

Contopoulos, G. 1979, in Photometry, Kinematics and Dynamics of Galaxies, ed. D.S. Evans (Austin: Dept. Astron. Univ. Texas at Austin), p. 425.

Contopoulos, G., and Mertzanides, C. 1977, Astr. Ap., 61, 477.

Contopoulos, G., and Papayannopoulos, Th. 1980, Astr. Ap., in press.

Crane, P. 1975, Ap.J., 197, 317.

de Vaucouleurs, G. 1963, Ap.J. Suppl., 8, 31.

de Vaucouleurs, G. 1974, in IAU Symposium No. 58, The Formation and Dynamics of Galaxies, ed. J.R. Shakeshaft (Boston: Reidel), p. 335.

de Vaucouleurs, G. 1975, Ap. J. Suppl., 29, 193.

de Vaucouleurs, G., and Buta, R.J. 1980, A.J., 85, 637.

de Vaucouleurs, G., de Vaucouleurs, A., and Corwin, H.G. 1976, Second Reference Catalogue of Bright Galaxies (Austin: Univ. of Texas Press).

Freeman, K.C. 1970, in IAU Symposium No. 38, The Spiral Structure of Our Galaxy, ed. W. Becker and G. Contopoulos (New York: Springer-Verlag), p. 351.

Freeman, K.C. 1975, in IAU Symposium No. 69, Dynamics of Stellar Systems, ed. A. Hayli (Boston: Reidel), p. 367.

Freeman, K.C. 1979, in Photometry, Kinematics and Dynamics of Galaxies, ed. D.S. Evans (Austin: Dept. Astron. Univ. Texas at Austin), p. 85.

Fried, J., and Illingworth, G. 1980, in preparation.

Hodge, P.W. 1974, Ap.J. Suppl., 27, 113.

Hohl, F. 1971, Ap.J., 168, 343.

Hohl, F., and Zang, T.A. 1979, A.J., 84, 585.

Huntley, J.M. 1980a, Ap.J., 238, 524.

Huntley, J.M. 1980b, private communication.

Illingworth, G. 1980, in Structure and Evolution of Normal Galaxies, ed. S.M. Fall and D. Lynden-Bell (Cambridge, U.K.: Cambridge Univ. Press), in press.

Iye, M. 1978, Pub. A.S. Japan, 30, 223.

Kalnajs, A.J. 1973, Proc. Astr. Soc. Australia, 2, 174.

King, I.R. 1975, in La Dynamique des Galaxies Spirales, ed. L. Weliachew (Paris: CNRS), p. 417.

Kormendy, J. 1979a, Ap.J., 227, 714.

Kormendy, J. 1979b, in Photometry, Kinematics and Dynamics of Galaxies, ed. D.S. Evans (Austin: Dept. Astron. Univ. Texas at Austin), p. 341.

Kormendy, J. 1980a, in Proceedings of ESO Workshop on Two Dimensional Photometry, ed. P. Crane and K. Kjär (Geneva: ESO), p. 191.

Kormendy, J. 1980b, series of papers in preparation.

Kormendy, J. 1980c, in preparation.

Kormendy, J., and Illingworth, G. 1980, in preparation.

Kormendy, J., and Koo, D. 1980, in preparation.

Kormendy, J., and Norman, C.A. 1979, Ap.J., 233, 539.

Kormendy, J., and Norman, C.A. 1980, in preparation.

Lindblad, B. 1956, Stockholms Obs. Ann., 19, No. 7.

Lynden-Bell, D. 1979, M.N.R.A.S., 187, 101.

Mebold, U., Goss, W.M., van Woerden, H., Hawarden, T.G., and Siegman, B. 1979, Astr. Ap., 74, 100.

Miller, R.H., and Smith, B.F. 1979a, Ap.J., 227, 407.

Miller, R.H., and Smith, B.F. 1979b, Ap.J., 227, 785.

Peterson, C.J., and Huntley, J.M. 1980, Ap.J., in press.

Peterson, C.J., Rubin, V.C., Ford, W.K., and Thonnard, N. 1978,

Ap.J., 219, 31.

Roberts, W.W. 1979a, in IAU Symposium No. 84, The Large-Scale Characteristics of the Galaxy, ed. W.B. Burton (Boston: Reidel), p. 175.

Roberts, W.W. 1979b, in Photometry Kinematics and Dynamics of Galaxies, ed. D.S. Evans (Austin: Dept. Astron. Univ. Texas at Austin), p. 461.

Sancisi, R., Allen, R.J., and Sullivan, W.T. 1979, Astr. Ap., 78, 217.

Sandage, A. 1961, Hubble Atlas of Galaxies (Washington, D.C.: Carnegie Institution of Washington).

Sandage, A., and Brucato, R. 1979, A.J., 84, 472.

Sanders, R.H. 1977, Ap.J., 217, 916.

Sanders, R.H., and Huntley, J.M. 1976, Ap.J., 209, 53.

Sanders, R.H., and Tubbs, A.D. 1980, Ap.J., 235, 803.

Sanders, R.H., and van Albada, T.S. 1979, M.N.R.A.S., 189, 791.

Schechter, P.L., and Gunn, J.E. 1979, Ap.J., 229, 472.

Schweizer, F. 1980, Ap.J., 237, 303.

Sellwood, J.A. 1980a, Astr. Ap., in press.

Sellwood, J.A. 1980b, preprint.

Shu, F.H., Stachnik, R.V., and Yost, J.C. 1971, Ap.J., 166, 465.

Simkin, S.M., Su, H.J., and Schwarz, M.P. 1980, Ap.J., 237, 404.

Toomre, A. 1977, Ann. Rev. Astr. Ap., 15, 437.

Tsikoudi, V. 1977, Pub. Dept. Astron. Univ. Texas, No. 10.

van den Bergh, S. 1960, Ap.J., 131, 215.

van der Kruit, P.C. 1979, Astr. Ap. Suppl., 38, 15.

Wray, J.D. 1979, in Photometry, Kinematics and Dynamics of Galaxies, ed. D.S. Evans (Austin: Dept. Astron. Univ. Texas at Austin), p. 311.

Zang, T.A., and Hohl, F. 1978, Ap.J., 226, 521.

WHAT AMPLIFIES THE SPIRALS?

Alar Toomre

Massachusetts Institute of Technology

It seems clear now that the spiral structure of galaxies is a complex riddle without any unique and tidy answer. At most, one might perhaps still hope to blame it all on differential rotation, one way or another. Beyond this quasi-platitude, however, we know that the logical paths diverge soon and deservedly toward such separate themes as global instabilities, stochastic spirals, and also the spectacular shock patterns that can arise in shearing gas disks when forced by bars, to cite just the three areas which have progressed most vigorously as of late.

Any reviewer of this knotty subject faces a breadth-vs-depth dilemma: Ought he really to try and grapple with several of these diverse topics? Or should one instead focus on just one or two? Not too long ago I took the first tack in an article (Toomre 1977) that ran to 40+ pages. Here I opt for the opposite course, and shall dwell mainly on a single topic from the wave mechanics of a disk galaxy. I wish to reacquaint you with a neat old phenomenon that I have lately taken to calling <u>swing amplification</u>.

At first, this choice of emphasis may strike you as unusually biased and narrow. I rather hope it does — for then I can teach you something — but I assure you it is neither. On the contrary, here seems to be a rare opportunity for a reviewer to point to an almost forgotten process and to urge "Wait, try this one again!". But why? For one thing, I believe this amplification will help us all, at long last, to understand (and to repair) many of the global instabilities that have plagued our subject for the past decade. And as you will see, this same powerful phenomenon also makes it astonishingly plausible that some of the finest "normal" spirals, like M51 and M81, may in fact be just beautiful tidal transients.

One extra reason makes this topic singularly fitting. Though not yet by that name, swing amplification in a fairly small-scale or local context was first noticed right here in old Cambridge, in late 1963, by Goldreich & Lynden-Bell (1965, hereinafter GLB). Now let me illustrate at once what it can do in the large, and why it deserves a lot more attention than it has received.

111

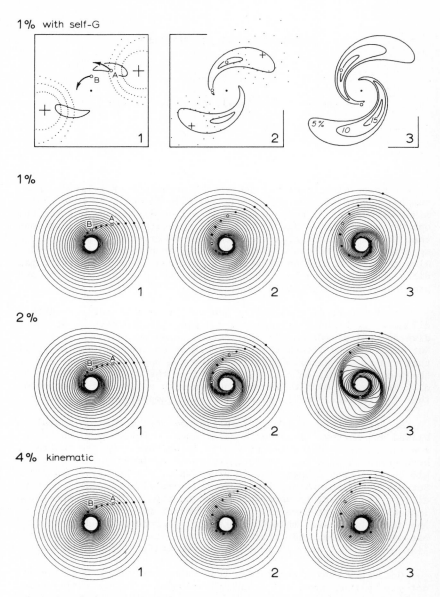

Fig. 1   Responses of Zang's V = const disk of stars to transient gravity forces from the imposed masses sketched at the upper left. The top row shows the excess densities among average stars. Other rows report the fates of various "cold" circles of test particles.

112

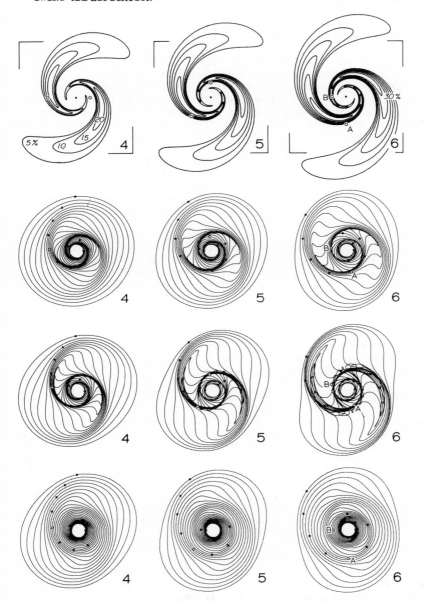

Fig. 1 (cont.) The top two rows presume that the maximum "tidal" force on particle A reached 1% of its central force; the third row supposes 2%. For comparison, the bottom row with its still fairly modest 4% forcing pretends that the stars do not respond at all.

113

ALAR TOOMRE

1 SOME REMARKABLE RESPONSES

Perhaps the first thing that needs to be said about Figs. 1 and 2 is that they display a vehemence that is not due to some computational blunder. Tom Zang and I have now checked them and their kin in too many independent ways to believe that — though it is true that we have probably stretched our small-amplitude stellar-dynamical analyses beyond their elastic limits in picturing these vigorous responses in the later time frames.

The second thing to be stressed is that these spiral responses also did not result from any "triggering" of some major two-armed instability. On the contrary, the unbounded disk model to which they refer seems quite stable now, at least against infinitesimal disturbances. This claim itself is unusual, and I think I had better elaborate on it.

The model employed here is basically the same as that studied by Zang (1976) in his thesis: a thin, infinite disk of stars with a Gaussian spread of radial velocities, all orbiting in well-mixed equilibrium in the V_o^2/r force field corresponding to the flattest imaginable law

$$V(r) = V_o = \text{const} \tag{1}$$

of circular motion. The only major revision from Zang's thesis is that we now postulate the density of mobile disk stars everywhere to be just one-half the amount

$$\mu(r) = V_o^2 / 2\pi G r \tag{2}$$

demanded by self-consistency. In other words, like many authors these days, we pretend that the missing half of the central force field comes from an allegedly rigid and fixed spheroidal component or halo. This last trick serves two purposes: (a) It stabilizes the one-armed unstable modes that were the only real nuisance in Zang's full-mass study. (b) It dramatizes that even a stable half-mass disk can retain a lot of vitality — despite the extra handicap that the rms random speed of the active stars was here chosen a factor $Q = 1.5$ times as large as still needed to curb the axisymmetric Jeans instabilities.

All of the disturbances pictured in Fig. 1 resulted from weak external "quadrupole" forces that were turned on and off smoothly but rapidly with time t (in the same units as our frame numbers) like $\exp(-t^2/2)$, while revolving counterclockwise with the same angular speed as the test particles in the outermost circles. The word quadrupole is correct: our imposed fuzzy masses included two similar negative "blobs" spaced $\pm 90°$ from the positive ones shown, to confine the excitation precisely to the $m = 2$ harmonic.

114

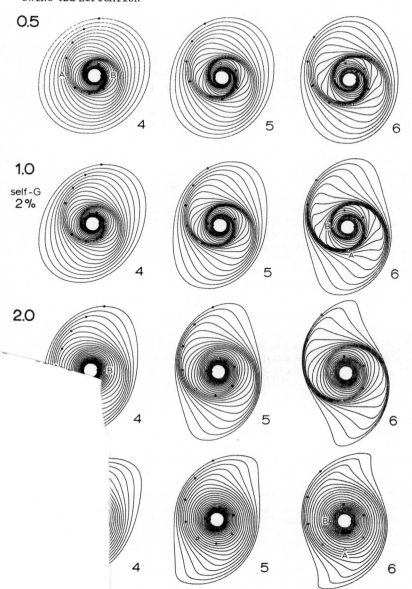

of V = const disk to masses imposed briefly at
w without rotation. In rows 2 and 4, the dura-
1 forces matches Fig. 1. In rows 1 and 3 it is
as large. The top three rows assume 2% maximum
A; the kinematic bottom row again supposes 4%.

115

As indicated by the top row of Fig. 1, these transient imposed forces — amounting even at their peak instant $t = 0$ to only about 1% of the galactocentric force on particle A, and to a yet smaller 0.25% upon particle B — soon yield an evolving spiral pattern of impressive severity among the typical disk stars.

Alas, such density patterns alone convey little feeling for what really happened here. That is why the second row of Fig. 1 repeats the message via the successive distorted shapes of what were initially just 25 concentric, logarithmically-spaced circles consisting of totally non-interacting test particles. These later displays borrow heavily, of course, from the beautiful diagrams of Kalnajs (1973), which showed how neatly a kinematic spiral can be assembled from a set of suitably-turned ellipses, each resembling one of Lindblad's old "dispersion orbits". Never mind that the present situation evolves, and is no longer kinematic. It remains instructive and simply delightful to observe here again — as indeed also from the whole Lin-Shu-Mark-Lau struggle to establish quasi-steady density waves — what grand patterns can emerge from individually rather dull and minor distortions.

But why should the cooperative responses in Figs. 1 and 2:

(a) end up so much stronger than the kinematic ones?
(b) appear to develop from the centers outward?
(c) have arms so well delineated by all the marked particles down almost to B at frame 4, for instance?

Or again, what $Q = 1.5$, WKBJ-type density wave would:

(a) amplify so fiercely?
(b) not begin to propagate inwards?
(c) let itself be sheared with nearly the full material speed?

A quick answer to all these questions is that our diagrams as yet exhibit little of the behavior expected of "ordinary" spiral waves, except in the interiors of the later frames — e.g., notice how particle B finally leaves its original arm between $t = 3$ and 4 , to become an arm-to-arm traveler.

What we are witnessing instead is genuine swing amplification: a strong cooperative effect that inhibits interarm travel and encourages gravitational bunching as long as the wave pattern in any given vicinity remains loose enough to continue to shear rapidly. Like all orbital clocks, this phenomenon simply runs its course sooner at the smaller radii — which is why we first notice these amplified waves down there, despite the smaller initial amplitudes. Similarly, the release of our test particles and stars alike from this kind of (evidently 9-to-12-month!) imprisonment in their home arms occurs soonest where the local year is shortest.

The obvious next question is: Why should stars conspire to act like that?

NSPIRACY

law in certain jurisdictions, the charge of
 at least three parties. And so it is here.
 results from a three-fold conspiracy between
self-gravity.

his, it helps to think small. Just as GLB did
is pretend that all wavelengths and other scales
iat we can plausibly confine our attention to
:ing patch from the disk of a galaxy. Within
:d (say) on radius r_o and longitude $\theta = \Omega_o t$ —
: the mean angular speed — let us presume that
lequately the linearized equations

$$_o^{A}{}_o x + f_x \tag{3}$$

$$f_y$$

all but not-quite-Cartesian radial and tangential

$$\tag{4}$$

center of reference. Here f_x, f_y are disturbance
pecified, and

$$\Omega/dr \big|_{r=r_o} \tag{5}$$

rt constant of differential rotation.

have, of course, had a long and honorable past.
was essentially from them (with the disturbance
)) that Oort and Lindblad in 1927 exhibited both
: flow

$$\tag{6}$$

:" <u>shaking</u>, with squared frequency

$$\Omega_o A_o \ , \tag{7}$$

ne three partners in our alleged conspiracy.

nething small but crucial. This local shear flow
 vibrations share the same sense in any normal
r speed $\Omega(r)$ decreases outward. Both types of
 direction opposite to Ω itself. It is precisely

this agreement, aided by a rough match between $2A_o$ and κ even in magnitude, that makes it possible for a wide-open pattern of would-be epicyclic vibrations to resonate a bit with the shear flow as the latter swings it around rapidly toward more and more trailing shapes. The only extra need is for interstellar communication — and this, of course, brings us not to any radio signals but to our third partner, the <u>self-gravity</u>.

To picture all three culprits in action, think of the situation sketched in Fig. 3. Somehow or other, let us suddenly create the disturbance motions suggested by the arrows in the leftmost frame, and now let them shear at the $V = const$ rate from before. (To be sure, the excitation in Figs. 1 and 2 was applied from angles more akin to $\gamma = 0$, but the present longer swing from $\gamma = -60^{\circ}$ to $+60^{\circ}$ is easier to visualize.) The key point to observe here is that, whereas the epicyclic motion with its Coriolis forces tries to curve any given star clockwise out of a striped zone marking the excess density, those stripes themselves keep turning in the same sense and thereby tend to frustrate the escape plans.

The delay involved is quite serious: In an axisymmetric case without self-gravity, a given particle would cease to contribute to the excess density already at a time such that $\kappa t = \pi$. Here, by contrast, the full swing from $\gamma = -60^{\circ}$ to $+60^{\circ}$ requires $\kappa t = 2\sqrt{2} \tan 60^{\circ}$ — or 56% longer — but even so in the last frame our

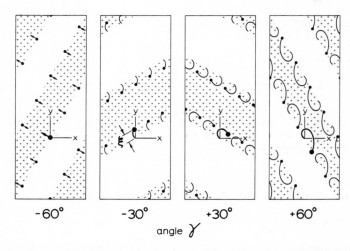

-60° $\qquad$ -30° $\qquad$ $+30^{\circ}$ $\qquad$ $+60^{\circ}$

angle γ

Fig. 3 Schematic of a shearing wavelet. In this imagined small patch cut from a larger disk, x points outward, y forward, and the angle γ reckons how much the striped pattern trails from the wide-open position. These four frames are spaced equally in time and include sheared epicyclic motions drawn assuming $V = const$.

typical star has not quite yet parted company with its neighbors.
The net result of all this kinematic lingering is that the average
star gets to contribute considerably more than its usual share
(roughly that 56% more) to the mass of any given shearing feature.
And if the effective mass of such an arm is boosted so unfairly,
so is its gravity and also its ability to attract yet other stars
toward itself in a kind of avalanche — or "sheared gravitational
instability", as GLB put it — while the rapid swing phase lasts.

This is about all I can offer by way of intuition. Now let me
buttress it with some mathematics.

The simplest measure of the strength of any given, spatially
sinusoidal wavelet is clearly neither x nor y. It is instead
the normal displacement

$$\xi = x \sin \gamma + y \cos \gamma \tag{8}$$

already sketched in Fig. 3. If our shearing patch were truly cold
— or had no stellar epicyclic motions whatsoever to begin with —
this ξ together with the instantaneous wave number

$$k = k_y \sec \gamma \tag{9}$$

and mean surface density μ_o would fully determine the perturbed
densities, and hence also the gravitational force

$$g = 2\pi G \mu_o k \xi \tag{10}$$

per unit mass exerted by the wavelet upon a particle remaining
near $x = y = 0$. This acceleration itself would always be orthogonal
to the wave troughs or crests, and so we could at last specify

$$f_x = g \sin \gamma$$
$$f_y = g \cos \gamma \quad . \tag{11}$$

At this point, if the wheel still needed inventing, it would
be cunning to pause and wonder whether, using forces (11), ξ by
itself might not be governed by some single equation of motion
more transparent than (8) and the pair (3). It is in fact so
governed, by the remarkable equation

$$\ddot{\xi} + \tilde{\kappa}^2(\gamma) \, \xi = g \quad , \tag{12}$$

with

$$\tilde{\kappa}^2(\gamma) = \kappa^2 - 8 \Omega_o A_o \cos^2 \gamma + 12 A_o^2 \cos^4 \gamma \quad , \tag{13}$$

that was in essence known already to GLB. (It is a lot easier to
verify than discover: start from (8) and $d(\tan \gamma)/dt = 2A_o$.)

119

ALAR TOOMRE

The beauty of this <u>GLB equation</u> — as it deserves to be called — is that it makes explicit, for each dimensionless shear rate

$$\Gamma = 2A_o / \Omega_o = - d \ln \Omega / d \ln r \tag{14}$$

and at each angle γ, just how much the usual epicyclic "spring rate" κ^2 is compromised by the near-resonance with the shear. An example of the reduced rate $\tilde{\kappa}^2(\gamma)$ is shown dotted in Fig. 4.

But why does one need such a spring at all? Won't it suffice to include lots of random motion and other pressure-like forces, such as are plainly needed to save our cold patch from devastation by Jeans instabilities at the shorter length scales? Here the old answer from 1964 seems well worth repeating: The stabilizing of even a smallish region of a disk against its self-gravity is very much of a <u>shared responsibility</u>, between pressure effects on the short side and the centrifugal or Coriolis forces on the long end. The former simply cannot help much — at least not until crosswise dimensions have been shrunk enough by the shear — if security is breached at large scales by any serious weakening of κ^2.

To estimate how much weakening can be tolerated, notice that (10) and (12) condense to

$$\ddot{\xi} + (\kappa^2 - 2\pi G \mu_o k) \xi = 0 \tag{15}$$

in the absence of shear. This means, of course, that wavelengths $\lambda = 2\pi/k$ in excess of

$$\lambda_{crit} = 4\pi^2 G \mu_o / \kappa^2 \tag{16}$$

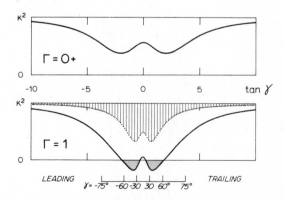

Fig. 4 Comparison of effective spring rates. The dotted curve gives $\tilde{\kappa}^2(\gamma)$ from (13). The solid curves, for $Q = 1.5$ and $X = 2$, show $S(\gamma)$ from (19), and include self-gravity and random motions.

120

would all be spared from Jeans instability even without recourse
to pressure effects. All the same, it would be a gross blunder to
dismiss the self-gravity of any wave of instantaneous length (say)
twice or even three times this critical value. On the contrary,
(15) reminds us also that such self-attraction would still depress
the effective κ^2 by either one-half or one-third.

Now return to the dotted curve in Fig. 4, and recall that our
finite shear rate $\Gamma = 1$ had by itself shrunk $\tilde{\kappa}^2$ to as little as
one-third of κ^2 at certain wide-open angles. Abbreviating the
ratio of unwrapped vs. critical wavelengths as

$$X = \lambda_y / \lambda_{crit} = 2\pi / k_y \lambda_{crit} \quad , \tag{17}$$

we then see at once why the choice $X = 2$ invites serious trouble:
as echoed by the solid curve, that extra one-half from the gravity
suffices to depress the overall stiffness $S(\gamma)$ briefly to quite
negative values near $\gamma = \pm 30°$. At smaller $X'es$, the gravity is
stronger and the danger yet greater — though the avalanching may
also stop sooner as pressure effects come into play. Conversely,
to ensure that $S(\gamma)$ never goes negative, the disk density μ_o has
to remain low enough (for given λ_y) to imply that X ≈ 3 or greater.
And that is essentially what it takes to kill swing amplification
when V = const .

I have not yet said anything very quantitative about the (here
fairly secondary) role of random motions of stars. Nor will I,
except briefly. For this, I have an alibi: Anyone really interested should now go and look up the old paper by Julian & Toomre
(1966, hereinafter JT). JT not only rediscovered for themselves
the similar local swing amplification in a shearing stellar sheet,
but they also discussed it extensively. Those results were more
complex, of course, but not essentially different from any above.
For instance, JT already thought it "especially interesting that
the greatest maximum [among several forced responses compared in
their Fig. 6] occurs for an unwrapped wavelength almost twice as
great as the axisymmetric λ_{crit} ".

To lend some visual feeling for what random motions add to the
picture, however, I concocted a hybrid for use in Figs. 4-7: Very
much in the spirit of the Lin-Shu (1966) "reduction factor" meant
for WKBJ density waves, I reduced the gravity claimed in (10) by
exactly the same complicated factor F which their dispersion
relation and that of Kalnajs (1965) advise for non-shearing waves.
With this simple but ad hoc modification, the GLB equation becomes

$$\ddot{\xi} + S(\gamma) \xi = 0 \quad , \tag{18}$$

where S abbreviates the whole combined spring rate

$$S(\gamma) = \tilde{\kappa}^2(\gamma) - 2\pi G \mu_o k_y \sec \gamma \cdot F(\gamma; Q, \ldots) \quad . \tag{19}$$

121

Hybrid though it is, I prefer this kind of an imitation to the gas pressure adopted by GLB, which would have yielded

$$S(\gamma) = \tilde{\kappa}^2(\gamma) - 2\pi G \mu_o k + c^2 k^2 \qquad (20)$$

with sound speed c . Such a make-believe gas tends to act rather too stiff and bouncy when more than just marginally stable, and in a full disk it can carry some quite unrealistic feedback signals via sound waves.

Two of the net spring rates (19) may be seen in Fig. 4, so to speak before and after the inclusion of the shear terms of GLB. The slow rise back toward κ^2 as $|\tan\gamma|$ increases and the true wavelengths become shorter and shorter is a familiar result from WKBJ theory, but the big "sag" of $S(\gamma)$ toward the middle is not. Instead, the basic WKBJ theory of Lin & Shu shifts there to its "long" branch — which is why the $\Gamma = 0+$ curve in Fig. 4 even rises a bit in the middle, as it so increases its frequency.

Enough now of spring rates. What do some numerical solutions of (18) actually look like? Figs. 5-6 furnish two samples, using the same parameters as employed not only in Fig. 4 but also for the global responses shown in Figs. 1-2. Fig. 5 refers to a half-swing from an impulsive start at $\gamma = 0$, whereas Fig. 6 concerns a full swing from very tight leading to very tight trailing angles.

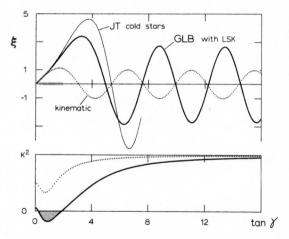

Fig. 5   Amplification during a half-swing, for $Q = 1.5$, $X = 2$ and $\Gamma = 1$. The heavy upper curve shows $\xi(t)$ inferred from (18) after an impulsive start, using the spring rate redrawn below. The dots refer to purely kinematic (or epicyclic) motions obtained without self-gravity. The remaining curve offers a more accurate stellar-dynamical result calculated in the manner of JT.

122

The impulsive excitation in Fig. 5 resembles those from Figs. 1 and 2 at least to the extent that any star 45° (of longitude) behind A or B would have felt the same sort of sudden forward pull as the present $x = y = 0$ particle from Fig. 3. Admittedly, those global examples also included immediate radial forces — such as on A and B themselves — but even there the tangential forces were probably more important by at least a factor 2.

The outcomes, at any rate, are strikingly similar: Especially from the JT-related curve which refers to the coldest subspecies, we see that the local half-swing amplification results in eventual oscillations with amplitudes about 4 times larger than kinematic. We observe also that the self-gravity delays severely the arrival of the first maximum, with the result that the major "arms" with gravity will appear more tightly wound by at least a factor 2, and denser by at least 8, than similar kinematic features. Moreover, the pitch angle of those first dynamical arms should correspond to $\tan\gamma \simeq 4$, or to inclinations $i = 90° - \gamma \simeq 15°$. Finally, Fig. 5 also suggests that the typical star will leave its fellows and pass through $\xi = 0$ at $\tan\gamma \simeq 5.4$ — which translates to about 10 months out of the local galactic year.

I would like to close with one additional item of hard-earned evidence that has persuaded Tom Zang and myself that his $V = const$ disk exhibits swing amplification rather than some rare tropical malady. In brief, we made it a point to examine carefully, for many <u>length ratios</u> X and for the three indicated values of the <u>speed ratio</u> Q, just how much amplification can be wrung out of.

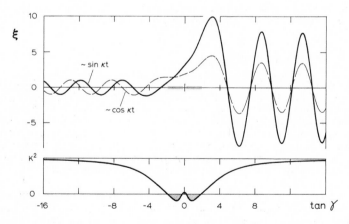

Fig. 6   Amplification from a full swing. The heavy curve above refers to a wavelet vibrating initially like $\xi = \sin\kappa t$. The broken curve started like $\cos\kappa t$. The effective spring rate is repeated below, again shaded where negative. Parameters are as before.

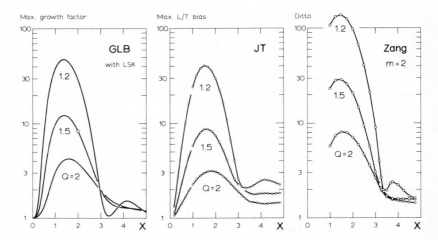

Fig. 7   Comparison of maximum full-swing amplification factors, between (a) GLB equation with LSK dispersion relation, (b) JT and (c) the V = const model subjected to two-armed perturbations.

the GLB/LSK, JT and Zang models under the various circumstances. The results are summarized in Fig. 7, where the single open circle in the leftmost panel reports, for instance, that we could have obtained a net growth by factor 8.38 in Fig. 6 upon selecting the most optimal arrival phase. Well, at least my confederate and I consider these plots decisively similar! The full disk indeed responds yet more vigorously than suggested by our local estimates, but it too seems to know that the big fun is over once X > 3.

3   FATE OF A LEADING WAVE PACKET

A leisurely next step might be to illustrate how the local shearing wavelets from GLB or JT can be superposed into packets of initially leading waves which then proceed to reflect/transmit and swing-amplify themselves right across a corotation region. For brevity, however, let us continue right away to Fig. 8, borrowed once more from the extensive joint work which Zang and I hope to publish in the Astrophys. J. during the next year or so. It shows a vivid example of the same process taking place globally.

This example speaks almost for itself. Its rapid growth in frames 3-6 indeed stems from our now-familiar cause. However, the subsequent decay in frames 7-9 is due only partly to group transport of trailing waves away from corotation (CR). Equally to be blamed is the well-known fact that any inner and outer Lindblad resonances (ILR and OLR), if present and unsaturated, simply love to gobble up all arriving density waves.

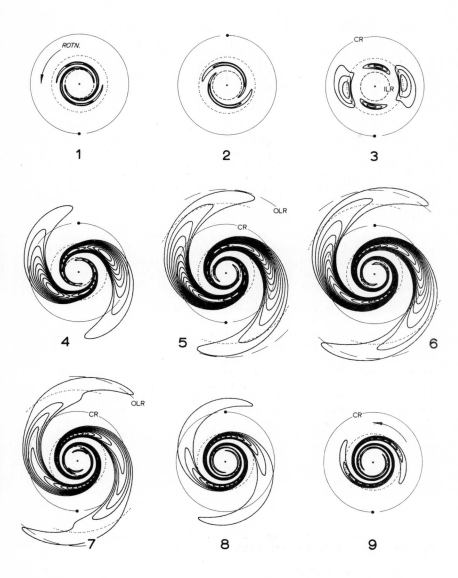

Fig. 8   From dust to ashes — fate of a packet of leading density waves in Zang's Q = 1.5 , X = 2 disk of stars.   Fractional excess densities in this evolving wave field are reported here at times separated by one-half of the nominal rotation period.   That is why these patterns hardly appear to rotate.

125

4 GLOBAL MODES OF A GAUSSIAN DISK

The brief flowering of the grand wave patterns in Fig. 8 evokes a series of follow-up questions: What if there had been no ILR to absorb these much-amplified trailing waves? Where would they have gone? And might some not have returned as fresh leading waves, to create further mirth or mischief?

Ironically, most of the answers — plus a brief mention of the "transient amplification" of GLB — can be found already on p. 909 of Toomre (1969), the paper which pointed out that WKBJ density waves possess a significant group velocity in the radial direction. As noted there, "even the ILR may be absent if the frequency of an assumed m = 2 disturbance is too large", in which case "the given wave packet must in some sense be reflected ... from the center" and "in the process, its character will presumably change from trailing to leading, and the sign of the group velocity should also reverse". Yet I quote all this hardly to boast. On the contrary, I very much failed to see then that already such reversed waves can serve as fresh grist for the swing-amplifying mill. It took several years more before any of us realized that all those simple pieces fit together neatly into a powerful feedback cycle.

Although still not in print, it was Bardeen in 1976 who first demonstrated clearly, from his global-mode studies of thin gaseous disks, that short leading waves returning from the central regions are indeed responsible for some of the most violent global instabilities. Zang and I soon corroborated this using waves reflected from sufficiently sharp "holes" carved into our V = const models. Today, however, let me retell the gist of those stories without leaning very directly on either Bardeen or Zang. Instead, I will now lean on Kalnajs, together with whom I have recently reexamined the two-armed global modes of the cold Gaussian disk that Erickson (1974; see also p.471 of Toomre 1977) first explored but could not stabilize despite his clever use of a "soft" close-range gravity.

The results of our study are summarized in Figs. 9-12, using a single fixed value

$$a = 0.25 \tag{21}$$

for the characteristic length of that gravity softening. As you can see, this Gaussian disk with surface density

$$\mu(r) = C \exp(-r^2/2) \tag{22}$$

remains unstable even in our hands. Like the gaseous Kuzmin disk studied by Aoki, Noguchi & Iye (1979), it certainly exhibits a long sequence of unstable spiral modes A, B, C, ... , the first few of which were known already to Erickson in the case of a simple disk subject to no extra forces from an imagined halo. The fiercest of these instabilities (with pattern speed $\Omega_p = 0.4153$ and growth rate $s = 0.2178$ in obvious units) is shown developing at the usual alarming pace in Fig. 9. This mode A would be referred to as the "bar mode" by most workers — especially after its growth rate is weakened and its planform made less spiral, as in Figs. 11-12, by locking up more and more of the disk mass for the sake of argument. It is also the kind of mode which, I dare say, practically every one of us has at some time guessed to be the one most closely analogous to the principal instability of the Maclaurin spheroids.

Plausible though it seems, the Maclaurin analogy is fallacious. Erickson's mode A is in truth just the first of a long series of swing-amplified modes, each of which is indebted also to the above-mentioned return of signals via the trailing → leading density waves. One can surmise this to a fair extent already from the great family resemblance of modes A, B, C, E, F in the density plots of Fig. 10, and from the shared trends of their pattern speeds and growth rates in Fig. 11 as the "active" disk mass is progressively reduced. But the real clincher in my opinion comes in Fig. 12, which reports the modal shapes for the case where only two-thirds

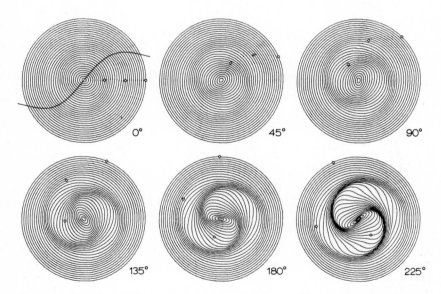

Fig. 9   Rapid growth of Erickson's (1974) dominant unstable mode A in the cold, $a = 0.25$ Gaussian disk. From frame to frame, this pattern turns exactly 45° counterclockwise, and it intensifies by a factor 1.51. Like the three marked particles from $r = 1, 2$ and 3, all constituents here orbited initially in concentric circles and shared the rotation curve superposed on the first frame.

of the disk density is presumed to remain mobile. Modes B, C, E, F
then show unmistakable interference patterns: The 90° spacing of
their successive density maxima argues eloquently for the presence
of trailing and leading waves of very similar wavelengths, whereas
their deep minima attest that even the amplitudes must be nearly
the same, especially toward the interior.

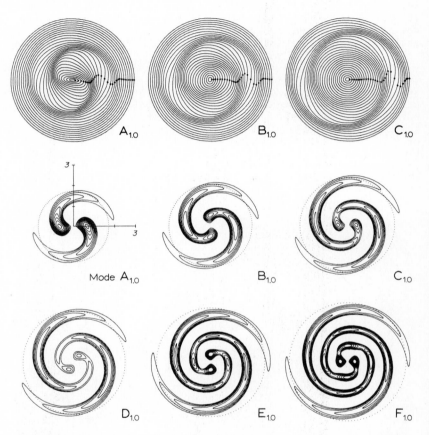

Fig. 10   Comparison of modes A-F for the full-mass Gaussian disk.
The top left entry repeats the 180° frame from Fig. 9; its marked
particles would have lain in a straight row in the absence of mode
A.   Immediately below it, and drawn to the same amplitude, scale
and orientation, are the perturbed density contours for this mode,
reckoned as 6, 12, 18, ... per cent of the central density.   The
similar matched pairs of diagrams for modes B and C presume 3, 6, 9,
... and 2, 4, 6, ... per cent, respectively.   Only the density
fields are shown for modes D-F.   The lightly dotted circles in the
middle and bottom rows mark the various corotation radii.

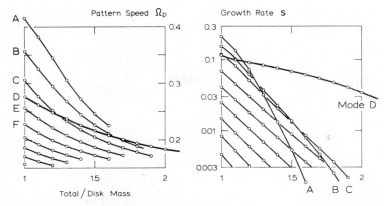

Pattern Speed Ω_p

Growth Rate s

Total / Disk Mass

Fig. 11   Pattern speeds Ω_p and growth rates s of various m = 2 modes of the a = 0.25 Gaussian disk, all obtained on the premise that only a <u>fraction</u> f of the disk density (22) responsible for the central force field is free to partake in these disturbances. The abscissae measure 1/f . For this disk, the highest Ω_p that would admit an ILR is 0.122; it lies well below the pattern speeds shown.   Axisymmetric stability requires only that a $\gtrsim$ 0.20 f .

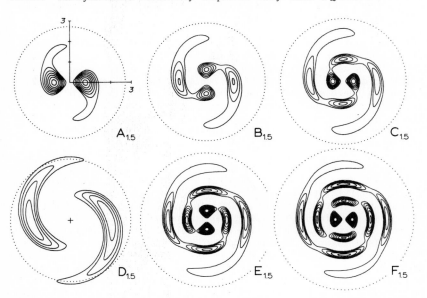

Fig. 12   Comparison of modes A-F for that Gaussian disk in which only 2/3 of the density remains "active". Their eigenfrequencies were reported at location 1.5 in Fig. 11. The corotation circles are again shown dotted; they have expanded markedly from Fig. 10.

As Bardeen recognized in 1976, such distinctive modal shapes will arise very naturally if individual leading waves or packets amplify at least a little when the galactic shear flow turns them into trailing waves near corotation, and if those trailing waves thereupon drift inward and eventually return as fresh leading waves to complete the cycle. From this viewpoint, it is also not surprising that each successive mode here, including mode A, seems to contain just one extra or one fewer wavelength.

Of course, one big technical question remains: Given what we know about the group-velocity drift rates, the extra time consumed by the rapid swinging, and the swing-amplification factors available from sources like Figs. 6-7, can we plausibly account for the growth rates actually obtained for modes A, B, C, E, F ... ? Happily, the answer seems to be yes — provided we allow for the fact that the unperturbed density (22) in the present model drops off quite rapidly near typical corotation radii $r \simeq 3$. This complication means that appreciable disturbance densities and hence forces can arise from mere radial displacements, rather than only from the compression of the disk material as we assumed earlier in (10). With such effects included, I now find that our crudely patched-together local theory accounts for something like 60 or 80 per cent of most of the growth rates in Fig. 11. Though hardly superb, this rough agreement leaves little doubt that we have grasped the essence of Erickson's modes A, B, C, E, F ...

One black sheep still needs to be dealt with. I am referring, of course, to the mode marked D in Figs. 10-12. As luck has it, the pattern speed (and even the growth rate) of this mode lands it smack amidst the swing-amplified modes in the full-mass Fig. 10. And it is there somewhat contaminated by the latter — as if only to confuse us! That mode D is a wolf in sheep's clothing becomes clear, however, once we weaken those rival modes in Figs. 11-12 by reducing the active disk mass. Its shape and hefty growth rate then point firmly to a different kind of animal.

What is mode D? It seems genuinely to be an <u>edge mode</u> which (a) arises only if the disk density drops off abruptly enough with radius, and yet (b) does <u>not</u> require any wave transport into or through the central regions. Kalnajs and I can support claim (a) with some experimental findings that any analogue of mode D occurs at most very weakly in the yet more soft-edged exponential disk — and it is altogether absent from Zang's $V = const$ disk — whereas it can be aroused to fresh fury by artificially truncating either of those disks in a smooth but sudden enough manner. We can also vouch for claim (b) with the little discovery that any "freezing" of our Gaussian disk inward of (say) $r = 1$ hardly alters the eigen-frequency of mode $D_{1.5}$ in the third digit. For these reasons and some others, Kalnajs suspects that mode D may be our true logical heritage from the sharp-edged Maclaurin spheroids — the heritage we failed to find earlier in the "bar" mode A.

130

5 WHAT DOES IT ALL MEAN?

As our little detour to mode D has just underscored, swing amplification does not explain every ache, pain, spiral arm and global mode of a shearing disk galaxy. Yet it probably explains quite a lot. Most remarkably, it even challenges head-on the two widely-held beliefs from recent years that

1. violent bar-making instabilities of disks can reasonably be prevented only by the presence of massive halos, and

2. at least the grandest-looking "normal" spirals, like M51 or M81 or NGC 5364, display long-lived wavelike patterns which resulted from milder forms of global instability.

5.1 T/W and all that

It seems to me that this first piece of folk wisdom rests on an innocent but shaky analogy, one whose weakness was long masked by something of a numerical fluke. Since about 1970, far too many of us (including myself!) have tended to presume intuitively that the most awesome instabilities met in various strongly shearing disks are essentially the same as either the dynamical or secular bar-makings exhibited by the non-shearing but sharp-edged Maclaurin spheroids. As implied by my earlier comments on "mode A" and its kin, I am convinced now that such intuition is false. Yet that we accepted it for so long is largely a compliment to the conjecture by Ostriker & Peebles (1973) that something as simple as their T/W criterion originating from the secular Maclaurin instabilities might also determine whether major bisymmetric instabilities can or cannot occur in a variety of other contexts. For a long time, the numerical evidence from shearing disks seemed more or less to concur. And so, lacking any real understanding of global modes like those in Figs. 9-12, this T/W guess came to be regarded not only as the best one available, but as a darn shrewd guess indeed.

In retrospect, it is ironic that most of those experimental "confirmations" of the Ostriker-Peebles conjecture can also be recognized as fine testimonials to swing amplification! Recall the two main impressions of the experimenters: (a) It seemed possible to stabilize a typical pure disk only by raising its gas pressure or stellar random motions to unbelievably large values, whereas (b) reducing the disk mass itself — by locking more and more of it into a dynamically inactive halo — eventually worked like a charm. In our terminology, those experiments seemed to be saying that the local speed ratio Q needs to equal at least 3 to ensure reasonable peace and quiet in a full-mass disk, whereas a general doubling or tripling of the length ratio X accomplishes the same goal almost regardless of Q (provided only Q > 1). These two conclusions, of course, resemble closely the trends summarized in Fig. 7c for swing amplification in Zang's V = const disk.

As far as I can tell, it really was only a fluke or coincidence that the possibilities for $m = 2$ swing amplification in shearing disks of stars turned out to be roughly equivalent to the energy ratios T/W in excess of 0.14 that Ostriker & Peebles knew to cause secular difficulties to the Maclaurin spheroids. If this seems incredible, recall that Bardeen (1975) cautioned that two of his shearing <u>gaseous</u> disks with $T/W \simeq 0.27$ seemed already to be stable. Also note that Aoki <u>et al</u>. (1979) reported a similar critical value of 0.23 for the gaseous Kuzmin disks. Why should "bar-making" be that much easier to switch off in a hot gaseous disk rather than a stellar system? Here again swing amplification unravels a mystery: As said before, dispersion relation (20) for gas gets excessively "stiff" (compared to LSK stars) when its local $Q \gg 1$.

All this hindsight applied to the Ostriker-Peebles conjecture could thus far still be denigrated as some sort of belated one-up-manship. However, their "necessary criterion" has a more serious flaw that probably spoils it even as a mnemonic. Let me explain.

Simply put, swing amplification never deserves all the credit or blame for any global mode which it helps to power. The other vital ingredient, as the striking contrast between Fig. 8 and Figs. 9-12 has already emphasized, is the return of fresh leading waves. A soft-centered disk like the Gaussian, which transmits vibrations readily through its interior, can thus tolerate hardly any swing amplification — even forgetting edge mode D. (In that spirit, a gently-rising rotation curve as observed in M33 indeed argues for substantial unseen mass.) Yet when such returning waves are impossible — as they are in the $V = const$ disk without any halo at all, provided $m > 1$ — a disk system can live happily but dangerously with the amplifier turned on, so to speak, but with its feedback wire no longer intact.

I say "no longer" because it seems only prudent to entertain the possibility that real disks, if they ever grew too massive within their spheroids or halos, may have been quite capable of tampering with that wire. After all, it takes only a small inward shift of (say) the lowest quartile of a disk mass to increase the angular speed markedly near the center and so create an ILR barrier to the transmission of future waves of a given pattern speed. We also know from work pioneered largely by Lynden-Bell & Kalnajs (1972) that any on-going spiral structure (including the flow of gas in bars) tends to export angular momentum outward from the deeper regions of a disk. This must indeed help them to contract. In short, I suspect that real disk galaxies have benefitted in the past from a great deal of self-regulated upward mobility of the densities, circular speeds and random motions alike in their deep interiors. Hence anyone who still insists that we must invariably crush the amplifier may well resemble the general who exhorts his troops to attack only the main strength of the enemy rather than his weak lines of communication!

132

5.2 Tidal waves after all?

If further proof is needed that the potency of swing amplification has lately astonished even this old advocate, let me supply it with these last few remarks and figures on the tidal theme.

Like Kormendy & Norman (1979) in their recent critique of the well-observed spirals, several of us including Tully (1974) and Piddington (1978) have voiced concern that the three exceptionally regular "grand design" spirals M51, M81 and NGC 5364 — which have served to motivate and illustrate so much of the Lin-Shu theory of quasi-steady density waves — all have major neighbors, and rather strange ones at that. Hardly anyone would (or should) dispute any longer that especially in those galaxies we must be witnessing density waves of some kind. But might their kind not be transient vibrations caused by recent close approaches of the companions and destined to fade away after just another revolution or two?

My own suspicions on this topic go back to my group velocity paper from 1969, and to the kinematic study of strongly interacting galaxies which my brother and I published in 1972. Unfortunately, those two studies together soon lured me into a serious conceptual blunder which continued to befuddle Toomre (1977). As can be seen from that review, I persuaded myself foolishly that "the first step [to driving a density wave tidally into the interior of an M51 or M81] would have had to be some major distortion of the outer disk or halo" and that only "this in turn may ... have induced a relatively transient spiral wave in the proper disk, traveling inward with the group velocity." In the case of M51, at least, the outer distortions are unmistakable. Yet the time needed there for wave propagation remains at least twice as long as the inferred interval since the encounter. Hence I concluded, sadly, that tidal origins of the main spiral structure did not look promising even in M51.

What I forgot in 1977 was any self-amplification, taking place more or less in situ, of the weak kinematic waves which an abrupt enough tidal forcing must excite directly even well inside a disk. How silly of me! Today it is only too clear from Figs. 1-2 that such wave growths in the interior — where all clocks run faster — can indeed outpace the tidal evolution of the exterior. This does not prove, of course, that M51 et al. are tidal spirals, but it sure establishes who was a needless pessimist.

To end this story of fresh tidal promise on a more affirmative note, Figs. 13-14 and 15 offer rough design studies for M81 and M51, respectively. In the former pair, two of the time sequences from Fig. 1 have been merely replotted, each as if viewed from 30° latitude; the culprit M82 was imagined to have passed at the top. To me, these later shapes already look hauntingly like the real M81, especially as studied from Westerbork. It is also a pleasure to find that the wiggles in the theoretical isovelocity contours

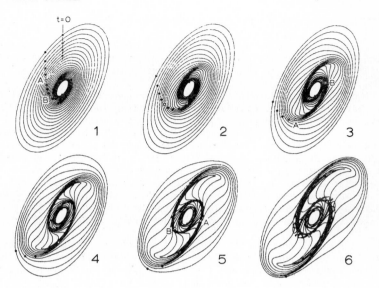

Fig. 13   A series of oblique views of the pictures from the third row of Fig. 1, again showing the approximate cold-gas response to that 2% tidal forcing. Shocks must evidently intervene by frame 5. This sequence presumes a 60° viewing inclination and -30° position angle; the peak tidal force was oriented exactly "north-south".

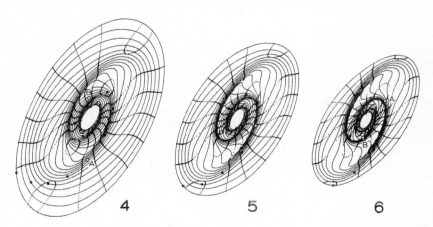

Fig. 14   Corresponding views of frames 4, 5 and 6 from the 1% row of Fig. 1.   These have here been scaled in ratio 15 : 12 : 10 , to stress the geometric similarity of the shapes though not of the amplitudes.   The dotted curves mark loci where the line-of-sight speeds equal 1.0, 0.75, ... , -1.0 times $V_o \cos 30°$.

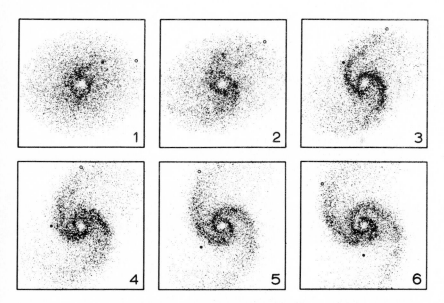

Fig. 15   Evolving spiral structure obtained by Zang from a 20,000 body simulation of the V = const disk. Time steps and the nominal disk parameters are as in Figs. 1-2. An m = 2 far-tide impulse at t = 0 provided a 20% velocity increment to the outermost particles. The two "clock" particles continue to orbit in circles.

shown in frames 4 and 5 of Fig. 14 — which assume about the same $0^{m}23$ density amplitude for stars as Schweizer (1976) reckoned from his photometry — resemble the 21-cm wiggles that Visser (1978) and others have attributed to a much more permanent wave.

The N-body experiment by Zang that is reported in Fig. 15 does not pretend to imitate M51 at all closely. It refers to a fairly abruptly truncated version of the V = const disk and remains mode- rately unstable, probably for that reason. Also, it was forced in a much too symmetric and somewhat too sudden manner. Nonetheless, Fig. 15 gives a glimpse at things to come. It certainly testifies again that it _is_ possible to manufacture a deep spiral structure before the outer boundary has distorted a great deal.

ACKNOWLEDGMENTS

In N + 1 ways I remain indebted to Tom Zang, a coauthor of this review in everything but name. Both of us are grateful to Jim Bardeen and Agris Kalnajs for much generous and wise advice. This work was helped also by grant MCS 78-04888 from the NSF.

REFERENCES

Aoki, S., Noguchi, M. & Iye, M. (1979). Global instability of polytropic gaseous disk galaxies with [Kuzmin's] density distribution. Publ. Astron. Soc. Japan, 31, pp. 737-74.

Bardeen, J.M. (1975). Global instabilities of disks. In Dynamics of Stellar Systems, ed. A. Hayli, pp. 297-320. Reidel.

Erickson, S.A. (1974). Vibrations and Instabilities of a Disk Galaxy with Modified Gravity. Ph.D. thesis, Mass. Inst. Technol., Cambridge, Mass. 179 pp.

Goldreich, P. & Lynden-Bell, D. (1965). Spiral arms as sheared gravitational instabilities. Mon. Not. Roy. Astron. Soc., 130, pp. 125-58 (= GLB).

Julian, W.H. & Toomre, A. (1966). Non-axisymmetric responses of differentially rotating disks of stars. Astrophys. J., 146, pp. 810-30 (= JT).

Kalnajs, A.J. (1965). The Stability of Highly Flattened Galaxies. Ph.D. thesis, Harvard Univ., Cambridge, Mass. 129 pp.

Kalnajs, A.J. (1973). Spiral structure viewed as a density wave. Proc. Astron. Soc. Australia, 2, pp. 174-7.

Kormendy, J. & Norman, C.A. (1979). Observational constraints on driving mechanisms ... Astrophys. J., 233, pp. 539-52.

Lin, C.C. & Shu, F.H. (1966). On the spiral structure of disk galaxies, II. Proc. Natl. Acad. Sci. USA, 55, pp. 229-34.

Lynden-Bell, D. & Kalnajs, A.J. (1972). On the generating mechanism of spiral structure. Mon. Not. Roy. Astron. Soc., 157, pp. 1-30.

Ostriker, J.P. & Peebles, P.J.E. (1973). A numerical study of the stability of flattened galaxies. Astrophys. J., 186, pp. 467-80.

Piddington, J.H. (1978). Origins of galactic spiral structures. Astrophys. Space Sci., 59, pp. 237-56.

Schweizer, F. (1976). Photometric studies of spiral structure, I. Astrophys. J. Suppl., 31, pp. 313-32.

Toomre, A. (1969). Group velocity of spiral waves in galactic disks. Astrophys. J., 158, pp. 899-913.

Toomre, A. (1977). Theories of spiral structure. Ann. Rev. Astron. Astrophys., 15, pp. 437-78.

Tully, R.B. (1974). The kinematics and dynamics of M51, III. Astrophys. J. Suppl., 27, pp. 449-71.

Visser, H.C.D. (1978). The Dynamics of the Spiral Galaxy M81. Ph.D. thesis, Univ. of Groningen, Netherlands. 144 pp.

Zang, T.A. (1976). The Stability of a Model Galaxy. Ph.D. thesis, Mass. Inst. Technol., Cambridge, Mass. 204 pp.

HI SPIRAL STRUCTURE IN M31 AND M33

J.E. Baldwin

Cavendish Laboratory, Cambridge

INTRODUCTION

Observations of the gaseous structure of spiral galaxies are important both for providing tests of dynamical theories of spiral structure and for elucidating the factors governing the rate of star formation in the arms. The spiral galaxies in the Local Group do not display the most elegant or dramatic spiral patterns but they do have recognisable arms and they do give birth to stars. Most importantly, we know more about M31 and M33 than we do about any other spiral galaxies and we shall continue to do so even as techniques improve.

The neutral atomic hydrogen in these galaxies has been studied for nearly 25 years with steadily improving resolution, so that the current best evidence only enlarges on a picture already somewhat familiar. But the recent work of Newton on M33 (Newton 1978, 1980) and Unwin on M31 (Unwin 1979, 1980a, b), using the Half Mile Telescope at Cambridge, provides the highest resolution data available on any galaxies other than our own and the Magellanic Clouds. The resolutions of 0.8 x 1.5 arcmin (160 x 300 pc) for M33 and 0.8 x 1.2 arcmin (160 x 240 pc) for M31 correspond roughly to typical cross-sections of the HI spiral arms in the Galaxy.

M33 - THE HI SPIRAL STRUCTURE

The distribution of HI integrated along the line of sight is shown in Fig. 1 as a radio photograph. The area visible corresponds almost exactly to the full extent of the optical image on the Palomar Sky Survey prints. The warps in the outer parts of the HI disk extend to the north-west and south-east of the ends of the major axis shown here. The distribution of HI is dominated by the irregular but narrow spiral features, which are only partially resolved by the beam. Near the major axis they have intrinsic widths of about 350 pc and are separated by approximately 2 kpc. They extend for typically 4 kpc before their identity

137

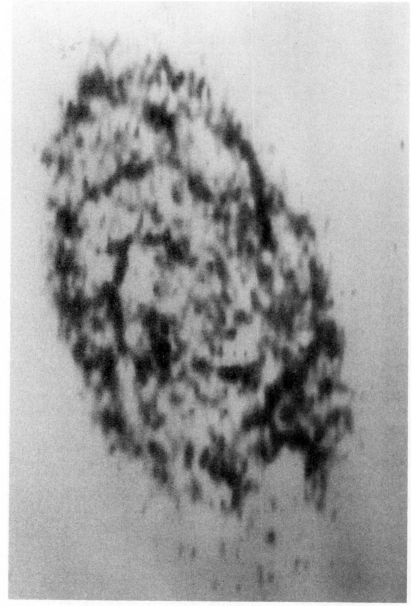

Fig. 1. Surface Density of HI to a radius of 6 kpc in M33.
The nucleus is marked by a cross.

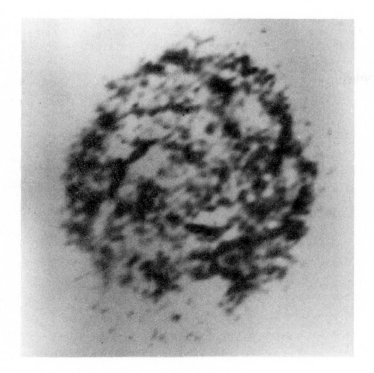

Fig. 2.   HI in M33 deprojected to face-on view.

becomes uncertain.

     The interarm region is not devoid of hydrogen but comprises
a roughly uniform background having a mean intensity of roughly one
fifth of the average peak intensity in the arms.   Thus the total
quantity of HI in the arms is just about equal to that in the
interarm region.   It is this ratio which must explain the lack of
certainty in detecting arms in earlier work of lower resolution.

     The velocity field over the whole of the area is extremely
regular, perhaps the most regular of any spiral known so far,
suggesting that the HI disk is flat and in circular rotation.   On
this assumption, taking the inclination to be 54°, the map of M33
deprojected to face-on view is illustrated in Fig. 2.   The beam
is, by chance, almost circular in this projection.   Within
2.5 kpc of the nucleus the HI structure is 2-armed and corresponds
to the well known optical arms.   Outside this radius segments of
perhaps 4 arms can be traced, all having very similar pitch angles

of about 27°.

A striking feature of the face-on view is that it is hard to distinguish which was the direction of the original major axis without resorting to identifying details. This implies that the thickness of the arms produces no discernible effects even near the minor axis. The apparent widths there of only $\sim$ 170 pc can contain no large contribution from the z thickness.

The hydrogen line profiles in Newton's work were all simple and could be characterized adequately by a peak brightness temperature, a mean radial velocity and a dispersion, σ, or width to half intensity, $v_{\frac{1}{2}}$. The profile widths are remarkably uniform over the whole of the disk with only a few points giving widths greater than 35 km/s (after correction for instrumental broadening). The mean rms dispersion in the line of sight was 9 ± 3 km/s and showed no significant variation with radius in the disk. The contributions to the observed σ^2 are roughly one third from σ_z^2 and two thirds from the components of σ^2 in the plane. The latter varies from σ_R^2 on the minor axis to σ_θ^2 on the major axis, so that there is some indication that σ_R and σ_θ do not differ.

If it is assumed that the velocity ellipsoid of the gas is a sphere, the thickness of the gas layer can be deduced once a good mass model of the disk is known. The rotation curve out to 6 kpc is well defined by the very regular velocity field mentioned above. Taking models in which all of the mass is concentrated in the disk, the observations are very closely approximated by a disk whose projected mass density falls exponentially with radius with the same exponent as that of the observed luminosity distribution, modified slightly depending on the assumptions regarding the z geometry of the mass distribution. The mass densities, ρ, at z = 0 can be used to calculate the thickness to half density of the HI layer

$$z_{\frac{1}{2}} = \frac{v_{\frac{1}{2}}}{(4\pi\rho G)^{\frac{1}{2}}}$$

The results for an axial ratio of the disk of 0.2 give an HI layer thickness of 250 pc at R = 2 kpc increasing steadily to 650 pc at R = 5 kpc. Surprisingly, the thickness at, say, R = 4 kpc is larger than the observed projected widths near the minor axis of $\sim$ 170 pc which are themselves largely accounted for by the width of the arm in the plane of the disk. So it is hard to account for this thin HI disk even if all of the mass is concentrated in the disk and it suggests that only a small fraction of the mass can be distributed in a halo, at least over radii in M33 out to 6 kpc.

M31 - THE HI SPIRAL STRUCTURE

Fig. 3 shows the distribution of HI in M31 out to a radius of 20 kpc in the disk as a radio photograph from the work of Unwin (1980 b). The ring of HI seen by Roberts (1966) and the arms detected by Emerson (1974) are seen to comprise a large number of long, narrow, spiral segments, many at least 10 kpc in length. Because of the high inclination (77°5) of M31, they cannot be traced reliably past the minor axis or at other places where they apparently intersect. Although there are large variations in their apparent thickness it is notable that the arms appear widest (1-2 kpc to half intensity points) where they cross the major axis and that the narrowest projected widths occur well away from the major axis. The most reasonable interpretation is that the arms have a greater width in the plane of the disk than in the z direction by a factor of about four. In M31, unlike M33, most of the HI (80-90 per cent) lies in the arms. The interarm projected density is often less than 1/10 of that in the arms whilst the ratio of arm separation to width is only 2-3. One cannot deduce the three-dimensional pattern of the arms from Fig. 3 uniquely but if it is viewed as a circular disk lying flat, there is a strong appearance of the SW (left hand) end being warped upwards on the page whilst the NE (right hand) is warped downwards. This suggestion is confirmed by the directions of the known warps in the fainter outer regions (75' < R < 150') not shown in this illustration (Newton & Emerson, 1977).

The narrowest projected widths of arms occur at R = 8-11 kpc in the SE and SW quadrants in Fig. 3 and correspond to $z_{\frac{1}{2}}$ thickness of these arms of 220-350 pc after correction for instrumental broadening. This value can be compared, as in M33, with a calculated thickness based on the velocity dispersion and a mass model of the M31 disk. Since M31 is viewed almost edge-on, the observed velocity dispersions at points on the major and minor axes are very nearly the tangential and radial components, σ_θ and σ_R, of the velocity dispersion. The observed values (Unwin 1979) are plotted in Fig. 4. The scatter of high values and the apparently larger values of σ_R are almost certainly due to blending of lines from arms having differing radial motions in M31 which are not large enough to result in double-peaked profiles near the minor axis. In other regions of M31 double peaked profiles are seen. The results are probably consistent with $\sigma_\theta = \sigma_R = 8.1 \pm 0.9$ km/s and with no change in this value with radius. It seems that there is very little variation in σ from galaxy to galaxy and almost no dependence on radius in a galaxy. With the further assumption that σ_z is also 8.1 km/s and the adoption of Emerson's (1976) mass models of the disk with an axial ratio of 0.1, the expected thickness of the HI layer increases steadily with radius from 200 pc at R = 8 kpc to 400 pc at R = 13 kpc and 600 pc at R = 17 kpc.

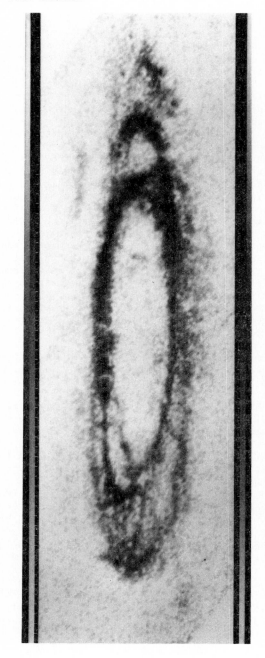

Fig. 3. The surface density of HI in M31 within about 20 kpc of the nucleus, which is marked by a cross. The NE end of the major axis is on the right and the near side of M31 is at the lower edge of the figure.

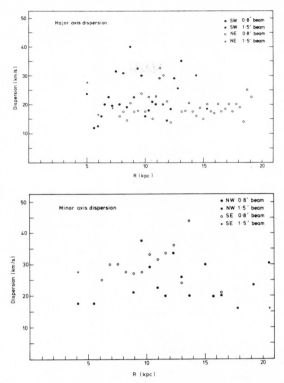

Fig. 4. Observed widths to half intensity of the H line profiles
along the major and minor axes of M31 as a function of
radius, corrected for instrumental broadening.

Comparison with the narrowest z thicknesses observed at
R = 7 kpc of 220 pc and 350 pc at R = 10 kpc suggests that the
agreement is surprisingly good in M31. This, however, ignores
the width of the arms in the plane. If they are as wide at these
points as on the major axis, most of the projected thicknesses are
accounted for by this alone, implying that the observed thickness
in z would be smaller than that calculated from the velocity dispersion. Agreement would be yet harder to reach if a substantial
contribution to the circular velocities were due to mass distributed in a halo at these radii.

There seem to be two possibilities for avoiding the conflict
in M33 and the possible one in M31:

(i) The stellar distribution in z is much thinner than is usually
assumed for disk galaxies

143

(ii) The true dispersions in the gas are smaller than those observed which are affected by large scale systematic motions such
as shocks within an arm.   This seems unlikely since the dispersions measured in the interarm regions are the same as
those in the arms.

The arm thicknesses deduced here lead to mean gas densities
in the arm which range from 3 atoms cm^{-3} at R = 2 kpc to 1 cm^{-3} at
R = 6 kpc in M33 and from 0.7 cm^{-3} at R = 10 kpc to 0.25 cm^{-3} at
R = 20 kpc in M31 and are respectively slightly higher and lower
than those deduced many years ago for similar averages in the
Galaxy.

KINEMATICS OF HI IN THE ARMS

M33 does not have a beautiful spiral pattern.   But it does
have spiral features and it offers some hope of detecting small
motions in them since the velocity field is very regular.   Within
R = 5 kpc, observations with a resolution of $\sim$ 1 kpc can be fitted
by circular motions in a flat disk with an accuracy better than
8 km/s.   If such a model is subtracted from the high resolution
observations, the residuals are again very small.   Over most of
the disk they are < ± 5 km/s and reach ± 10 km/s in only one or
two places.   In several cases, well defined segments of spiral
arms have no motions > 5 km/s (corrected for inclination) associated
with them.

How large are such residuals expected to be?   Since the
spiral structure shows no grand design, we must take a narrow view
and ask that question about a very small segment of an arm.
Extreme possibilities are

(i)   An arm that winds up by differential rotation quite freely and
rapidly.   In this case the excess density in the arm provides
very little gravitational effect and there need be only minute
peculiar motions associated with it.   The life-time of one
segment is short but other segments may perhaps be regenerated.

(ii) The arm pattern is very long-lived.   In this case there must
be extra radial and tangential motions in addition to pure
circular motion to maintain the arm rotating with some pattern
speed V_p at radius R.   For an arm with surface density much
larger than the interarm region and with pitch angle t, the
gas flow through the arm has a radial component

$$v_r = (V_c - V_p) \tan t \simeq 0.5(V_c - V_p) \text{ for M33.}$$

It is difficult to choose any single angular speed for the pattern
which will keep $(V_c - V_p)$ adequately small over the whole disk and
to maintain this explanation of the spiral structure one can only

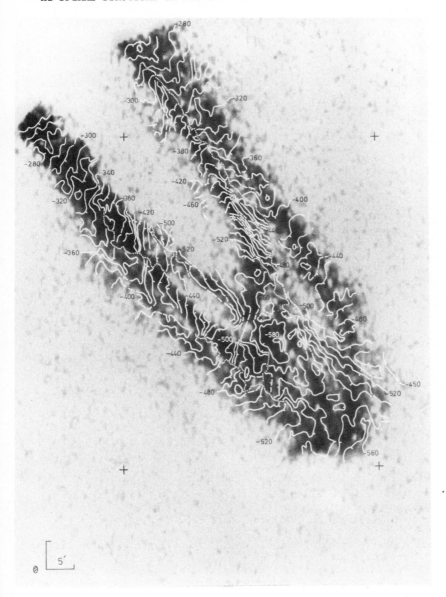

Fig. 5. The radial velocity field of the southern half of M31 superposed on the total HI map. The beam size is marked in the lower left-hand corner. The contours are at intervals of 10 km/s. (Unwin, 1979).

145

resort to the unsatisfactory supposition that V_p varies in such a manner as to keep $(V_c - V_p)$ small.

Overall it seems probable that M33 approximates to (i) rather than (ii). It would be of great interest to know what the peculiar motions associated with Toomre's arms are. My expectation from his pictures is that they are small (indeed, I would not wish to suggest, even obliquely, that he indulges in vigorous hand-waving) unless the density in the arm becomes very high, in which case his linear analysis would also need modification.

The velocity field in M31 is much more difficult to interpret. Fig. 5 shows the observed radial velocity field of the southern half of the galaxy. Unfamiliar features are the fine dotted lines which mark discontinuities in the plotted contours. They do not indicate true discontinuities, only that in regions where the profiles are double peaked this field has been plotted from the more negative radial velocity peak, provided that its intensity exceeds two to three times the rms noise. This procedure leads inevitably to sudden changes in the peak being fitted and to apparent discontinuities in radial velocity as one moves across the map. Similar diagrams can be prepared for the more positive velocity peaks and both have to be taken into account in considering the overall kinematics.

From this feature alone it is clear that residuals must occur when any model based on circular motions in a flat disk is subtracted from the observations. Even with the best-fitting models of this type the residuals are often as large as ± 20 km/s and extend over long continuous bands, roughly parallel to the spiral features. Such large scale residuals may be accounted for in at least three ways.

(i) An incorrect assumed inclination for the arm. The effect is not to change the value of the radial velocity perceptibly (M31 is nearly edge-on) but to change the apparent position of a feature in some assumed galactic plane. The effect should go to zero near the minor axis and be small near the major axis. Elsewhere a change of only 1° in inclination can give residuals of at least 8 km/s. Several arms may have contributions to their residuals from this cause.

(ii) Radial motions of a whole arm. These might give the largest residuals on the minor axis and would certainly not be expected to go through zero there. The strongest inner arm on the northwest side shows residuals of -20 to -30 km/s of this kind.

(iii) Shear of velocity across an arm which differs from the model velocity field. One clear example is the outermost strong arm on the north-west side where the iso-velocity lines run almost parallel to the minor axis for a length of about 10 kpc of the arm. This corresponds to rigid body rotation of the disk and would give long lifetimes for the arm segments. They have been discussed by Emerson (1976) and the present work of Unwin confirms them at higher resolution. It is striking that, whilst some arms show this behaviour, others do not and fit normal differential rotation. These differences from arm to arm in the trends of velocity across the arm are easily distinguishable from the effects in (i) and (ii).

The analysis of the velocity field is still in progress. We do not yet know whether it will be possible to separate these effects clearly.

REFERENCES

Emerson, D.T.E. (1974). Mon. Not. R. astr. Soc., 169, 607.
Emerson, D.T.E. (1976). Mon. Not. R. astr. Soc., 176, 321.
Newton, K. (1978). Ph.D. thesis, Cambridge University.
Newton, K. (1980). Mon. Not. R. astr. Soc., 190, 689.
Newton, K. & Emerson, D.T.E. (1977). Mon. Not. R. astr. Soc., 181, 573.
Roberts, M.S. (1966). Astrophys. J., 144, 639.
Unwin, S.C. (1979). Ph.D. thesis, Cambridge University.
Unwin, S.C. (1980a). Mon. Not. R. astr. Soc., 190, 551.
Unwin, S.C. (1980b). Mon. Not. R. astr. Soc., 192, 243.

NEUTRAL HYDROGEN IN GALAXIES

R. Sancisi

Kapteyn Astronomical Institute, Groningen

1. INTRODUCTION

Radio observations of neutral hydrogen gas in the Milky Way and in several dozen nearby galaxies have revealed the distribution and motions of the gas in these systems and additionally have been used to determine their total mass distributions and overall dynamical properties. The main results of such studies are reviewed and discussed in a number of articles (Roberts, 1975a; Burton, 1976; van Woerden, 1977; van der Kruit & Allen, 1978; Faber & Gallagher, 1979) and in the proceedings of recent conferences and symposia (IAU Symp. No.77 on "Structure and Properties of Nearby Galaxies", No. 84 on "The Large-Scale Characteristics of the Galaxy", and the conference on "Photometry, Kinematics and Dynamics of Galaxies", Austin, August 1979).

In this chapter some of the more recent work on the large-scale properties of the HI gas in elliptical, SO and spiral galaxies and in the neighbourhood of these galaxies will be discussed. These observations have revealed new aspects of the structure and dynamics of galaxies and constitute new evidence which may have a bearing on their formation and evolution.

2. ELLIPTICAL GALAXIES

Observations of elliptical (E) galaxies have shown that they contain little neutral hydrogen gas. The upper limits to the mass of HI in ellipticals obtained from 21-cm surveys, are about 10^7-10^8 solar masses (Knapp, Kerr & Williams, 1978; Gouguenheim, 1979; Sanders, 1980). These correspond to a fraction of the total mass less than 0.1% and to HI mass-to-luminosity ratios less than 0.001 to 0.05. These results are consistent, on the one hand, with the optical appearance of most ellipticals (lack of young stars and dust) but, on the other hand, are surprising in view of the substantial amounts of interstellar gas expected from the mass loss (typically 0.1 to 1 $M_\odot$ yr^{-1}, Rose and Tinsley, 1974) which is supposed to be shed from the evolved stars of E systems. This disagreement has led to the speculation that efficient removal mechanisms must exist for the gas in all ellipticals. At the same time, in order to explain the presence of HI in those few systems where it has been detected, recent infall is often advocated (cf. for instance Knapp et al., 1978). Detailed investigation of the spatial structure and kinematics of the HI in ellipticals is clearly important for the question of its origin; is it primordial material left over from the time of formation, is it material shed by stars within the galaxy or is it captured gas?

Two E galaxies have been studied in detail using the Westerbork Synthesis Radio Telescope; NGC 4278 and NGC 1052. NGC 4278 has about 2.5×10^8 $M_\odot$ of HI and $M_H/L_{pg} = 0.05$, about a factor 5 less than in Sb–Sc galaxies. The recent observations by Raimond, Faber, Gallagher & Knapp (1980) at Westerbork show that the HI is all within the Holmberg diameter (6 arcmin) down to the detection limit of about 1×10^{19} cm^{-2}. It is distributed in an apparent disk with some large-scale structure, possibly suggesting spiral arms, and it exhibits a depression in the central, optically bright region. Its velocity field shows large-scale regularity and the same overall pattern found in spiral galaxies, where it is usually interpreted as due to differential rotation. Some indications of a distortion or skewness of major and minor axis are also present, similar to those found in some spiral galaxies and ascribed to warping or "oval" distortions (Bosma, 1978). In the case of NGC 4278, where the line of nodes is not known from the optical picture and is not at all obvious from the shape of the HI disk, one cannot of course rule out the possibility that radial motions in the HI disk (inflow or outflow) may play a significant role. For the case of pure circular motion, the derived rotation curve rises steeply near the centre and remains roughly flat over the whole disk, as in many spiral galaxies. The maximum rotation velocity is 180/sin i km/s. For a possible range of inclination angles between 30 and 60 degrees the rotation velocities would be in the range 210 to 360 km/s, i.e. as large as or even larger than in massive Sb and Sc galaxies. The stellar component rotates much more slowly, with a maximum velocity of $\sim$ 50 km/s (Schechter & Gunn, 1979).

On the whole these HI results suggest a disk with properties similar to those of spiral galaxies, except for its low HI mass and surface brightness. However, comparison of the HI results with the optical picture indicates a potential problem: the HI kinematical major axis is rotated by about 40 degrees with respect to the major axis of the optical image (PA $\sim$ 20° for the optical and $\sim$ 60° for the HI). This difference may be reconcilable; the HI measures refer to the outer regions of NGC 4278, whereas the optical axis is defined at much smaller radii. Moreover the HI major and minor axes are derived only from the kinematics of the gas and may therefore be affected by the distortion or warping mentioned above.

The observations of NGC 1052 (Ekers & Goss, private communication) give detailed information on the HI only in the direction of right ascension. The results are very similar to, and consistent with, those obtained for NGC 4278; the HI is all within the Holmberg diameter and appears to be in well-ordered motion.

In conclusion the HI in the two ellipticals studied in detail seems to be in ordinary, probably rotating disks. Clearly, part of the evidence from HI observations (Knapp et al., 1978), which seemed to favour the hypothesis of a very recent infall of gas, no longer exists. Moreover, such rotating gaseous disks may be fairly common in elliptical galaxies, as dust lanes similar to that in NGC 5128 are seen in many ellipticals (cf. e.g. Kotanyi & Ekers, 1979).

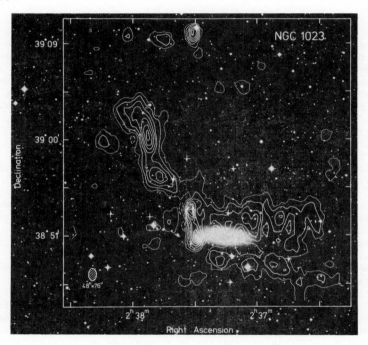

Fig. 1. Map showing the HI distribution in the field of the S0 galaxy NGC 1023 (Arp 135) (Sancisi et al., in preparation). The contour values are −3.4 (dashed), 3.4, 10, 20, 30, 40, 50, 60, 70x10^{19} cm^{-2}.

3. LENTICULAR (S0) GALAXIES

Galaxies of type S0 have the disk shape of spirals but contain

only an old stellar population similar to that in ellipticals
(Sandage, 1961). The absence of spiral arms, young stars and HII
regions has led to the suggestion that SO galaxies must be gas-
free systems. Contrary to this expectation the surveys of neutral
hydrogen in these systems carried out during the last decade (see
the review by van Woerden, 1977) have revealed a wide range of HI
content; the hydrogen mass to blue luminosity ratio, M_H/L_B varies
from values as large as 0.5, as found in late-type spirals, to
values lower than 0.01, typical of ellipticals. It is possible
that this result can be ascribed partly to misclassifications. On
the whole, however, the present observational evidence seems to
indicate that some morphologically normal SO's are as rich in
hydrogen as Sb or later spirals, whereas the majority have no
significant amount of HI; the peculiar objects tend to be richer
in gas.

A few of the SO galaxies already detected in the HI surveys
with single-dish radio telescopes have recently been studied in
more detail with aperture synthesis techniques (see Shostak, van
Woerden & Schwarz, 1979). In NGC 4203 the density distribution and
kinematics indicate a large-scale regularity similar to that found
in normal, later-type spirals. Part of the gas in NGC 4203 lies
within the bright optical disk, but most of it is found outside,
extending to almost 2.5 times the "Holmberg" size (van Woerden,
Gallagher & Schwarz, in preparation). The total mass of HI
($\sim 1\times10^9$ $M_\odot$) and the value of M_H/L_B (~ 0.1) are not
much less than in Sb galaxies but the HI surface brightness is
considerably lower, the gas being thinly spread over a very large
area.

The density distribution and kinematics of HI in NGC 1023 and
NGC 4694 are considerably more complex. In NGC 1023 (Fig. 1) the
majority of the gas ($\sim 1\times10^9$ $M_\odot$) is located around the system
and exhibits large-scale non-circular motions; another component
($\sim 0.5\times10^9$ $M_\odot$) is concentrated in a tail extending to one side
of the galaxy and out of its plane (Hart, Davies & Johnson, 1980;
Sancisi, Davies, Hart & van Woerden, in preparation). This
peculiar structure and kinematics, along with the presence of a
dwarf companion, suggests that a recent encounter with a gas-rich,
late-type galaxy may have led to a tidal disruption of the latter.
In this picture the optical companion is destined to fall toward
the centre and disappear shortly as a consequence of dynamical
friction. All the gas or most of it may arrange itself in circular
differential motion in the disk of NGC 1023 within about one
orbital period ($\sim 1\times10^9$ yrs in the outer parts). At present, the
optical morphology of NGC 1023 leaves little doubt about its being
of SO type, but as a result of the acquisition of a substantial
amount of gas, the system may develop into a later-type spiral.

This interpretation of NGC 1023 finds some support in the
observations of NGC 4694 (Shostak et al., 1979) where such a

merging of two systems may also be taking place and may perhaps
have reached an even more advanced stage. In this system, the HI
is also partly concentrated in a tail (Schwarz & van Woerden,
private communication) extending to one side of the galaxy, with
no optical counterpart. The remainder of the gas is found
concentrated to the centre of NGC 4694. An alternative explanation
for the distribution and motion of the gas might be found in
stripping by intergalactic gas; NGC 4694 lies in the outer parts
of the Virgo Cluster. The classification of this galaxy is
somewhat more uncertain than that of NGC 1023: its irregular
appearance may suggest an I0 type.

Yet more peculiar, but perhaps also falling in the same category, is the SOp galaxy NGC 2685, the "Spindle" (Hubble Atlas
p.7); recent accretion of gas has been advocated by Shane (1980)
in order to account for the remarkable kinematics of HI and stars.

In conclusion, a variety of relative gas content and HI
morphologies is observed in SO galaxies. This may reflect a wide
range of evolutionary stages in these objects, from almost complete depletion of gas to recent or current accretion. It is clear
that detailed observations of the distributions and motions of gas
in additional objects are needed to substantiate this hypothesis.

4. SPIRAL GALAXIES

4.1 Central regions, spiral arms, bars

The distribution of neutral hydrogen in ordinary spiral
galaxies is characterized by a deficiency in the central regions
and by spiral arm structures over the main part of the disk. The
kinematics is dominated by differential rotation over the whole
disk, but deviations from circular motions are often found in the
central regions, in arms, and in bars.

The central deficiency of HI generally obtains within a radius
of 3 to 5 Kpc, as in our Galaxy, and seems to be more pronounced
in early-type systems such as the Sab galaxy M81 (Rots, 1975).
Bosma (1978) has pointed out that in galaxies with conspicuous
bulges no HI emission is detected in the central region. But some
deficiency is also found in later types like the Sc system M101
(Bosma, Goss & Allen, 1980). The depth of this HI "hole" and its
radial extent are poorly known except in a few large, nearby
systems. The main observational limitations are lack of spatial
resolution and insufficient sensitivity. The latter is more
serious in the central region than in the outer parts because of
the larger velocity spread. HI absorption against a central,
extended radio continuum source can also be a cause for uncertainty. The origin of the depression is still a matter of debate.
In our own Galaxy, two additional phenomena may be germaine to the

understanding of the apparently missing HI gas. These are: (i) the presence of a substantial amount of gas in the form of molecular clouds in the central 1 Kpc region and (ii) the large non-circular motions of the HI within 3 Kpc of the centre (see discussion by Sanders, 1979).

In the spiral galaxies of large angular size, such as M51, M81, M101 (see Fig. 2) and IC 342, the neutral gas appears to be mainly concentrated in spiral arms. In general there is good spatial agreement with the optical arms; the HI coincides with the narrow dust lanes. The observed arm-interarm contrast varies from values of about 2, as found in M81 (Rots, 1975), to values as large as 10 found in M31 (Unwin, 1980). Systematic deviations from circular motion of order 10 km/s are seen in the gas kinematics near the arms in M81 and M51. In some galaxies a more amorphous or irregular distribution is observed. This is partly due to insufficient angular resolution. There are also galaxies like NGC 4736 (Bosma, van der Hulst & Sullivan, 1977) where part of the gas is found coincident with faint outer rings.

In barred spirals the distribution and kinematics of HI are less well known; the angular resolution and sensitivity of the observations are generally inadequate for detailed study. The SBb galaxy NGC 5383 is one of the very few extensively studied objects

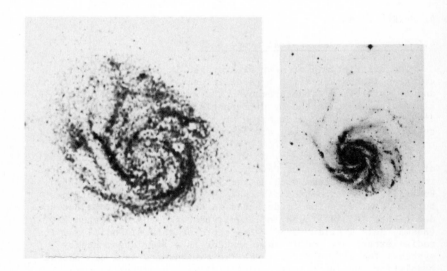

Fig. 2. M101. Left. Photographic representation ("radiograph") of the total HI surface density (Allen & Goss, 1979). Right. Blue photograph from Arp's Atlas of Peculiar Galaxies. Scale same as left.

(Sancisi, Allen & Sullivan, 1979). The main results for this object are: (i) the HI is present in the region of bar and arms, and
also shows a concentration in the central area, (ii) in the region
of the bar the radial velocities deviate from pure circular motion
($\sim$ 100 km/s); the isovelocity contours are skewed parallel to the
bar rather than parallel to the minor axis of the galaxy. Bosma
(1978) and Huntley (1978) have shown that this skewing of iso-
velocity contours can be produced by gas motion on highly elliptical streamlines which are elongated in the sense of the bar. At
present more detailed information on the kinematics of the gas in
the region of nucleus and bar comes from optical spectroscopy,
whereas the HI data provide the overall HI distribution and
velocity field.

4.2 Radial extent of HI; very large HI disks

The neutral hydrogen in spiral galaxies is generally found to
extend farther out than the optical image on the Palomar Sky Survey Prints. In our Galaxy (cf. Burton, 1976) the radial distribution and extent of atomic hydrogen is quite different from other
population I constituents such as young stars, ionized hydrogen,
molecules, cosmic rays and magnetic fields (as traced by synchrotron radiation). The HI extends from 4 to 14 kpc with little
density variation, whereas the other population I material lies in
a ring between 4 and 8 kpc. The situation in other spiral galaxies
is probably similar; the HI certainly extends farther out than HII
regions, young stars and radio continuum emission. Its surface
density decreases even more slowly (cf. Bosma, 1978) than the
total stellar luminosity, which drops exponentially or even cuts
off at some radius (cf. van der Kruit & Searle, 1980). Moreover
there is a suggestion that at large radii HI and "dark" mass
(inferred from rotation curves) may decrease similarly (Bosma,
1978) and may therefore be coextensive. Various questions which
arise are:
1) how far out does the HI extend in isolated spiral galaxies.
2) is the gas distributed in an extended disk or in a halo,
   smoothly or in discrete clouds.
3) is the HI differentially rotating or do random and non-
   circular motions dominate in the outer regions.
The answers to these questions have a bearing on the nature of
galactic halos, the properties of the intergalactic medium and the
origin of absorption lines in QSO's.

A convenient way of defining the HI size is to specify it at a
given "isophotal" level. This has been done by Bosma (1978) for
about 20 galaxies at the level of 1.8×10^{20} atoms cm^{-2}. The ratio
of the HI radius at this level to the Holmberg radius (at 26.5
photographic mag arcsec^{-2}; Holmberg, 1958) is generally less than
1.5, independent of morphological type. In the case of the giant
edge-on Sb galaxy NGC 4565, for example, the Westerbork HI map
(Fig. 3) shows that the HI at the level of $\sim 1 \times 10^{20}$ cm^{-2} extends

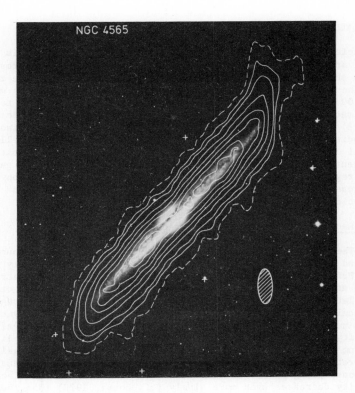

Fig. 3. Map of the HI in the edge-on Sb galaxy NGC 4565 (Sancisi, in preparation). The contour values are 1 (dashed), 3, 5, 10, 15, 20, 25, 30, 35×10^{20} cm^{-2}. The beam width at half power 51"x116" is shown by the hatched ellipse.

barely beyond the Holmberg radius (cf. Sancisi, 1978). Only in 3 of the isolated galaxies in Bosma's list does the HI at such high column density extend beyond 2 Holmberg radii. One of these, NGC 2841, is shown in Fig. 4. The question is whether beyond these still fairly dense regions the HI cuts off sharply or extends farther out at lower column densities. A recent survey (Briggs, Wolfe, Krumm & Salpeter, 1980) using the Arecibo radio telescope to search for extended HI envelopes surrounding spiral galaxies has shown that column densities $N_{HI} \gtrsim 3 \times 10^{18}$ cm^{-2} do not generally occur in isolated galaxies at 2 or 3 Holmberg radii. Only one (NGC 628) out of seven isolated galaxies shows emission at about 2.7 R_{HO} at 3×10^{19} cm^{-2}. In addition to this and the three galaxies of Bosma's list, there are a few more objects, also isolated, which were noticed for their unusually large HI extensions. Table 1 gives some of the main properties of these systems. In general more than 40%, and in some cases even more than 70% or 80%, of the total hydrogen mass lies outside the Holm-

156

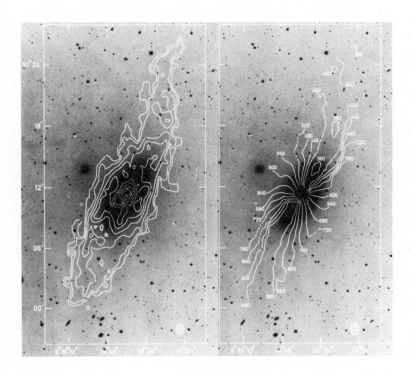

Fig. 4. NGC 2841. <u>Left.</u> Map of the HI density distribution
superposed on a III aJ print. The contour values are 1.05, 2.1,
4.2, 6.3, 8.4 and 10.5×10^{20} cm^{-2}. <u>Right.</u> Map of the radial
velocities. The numbers indicate the heliocentric velocity in
km s^{-1} (Bosma, 1978).

berg radius. The published HI maps of these objects (see e.g.
Mrk348 in Fig. 5, or M101) look very smooth, mainly because of the
low angular resolution. The gas might all be in dense clumps. The
objects listed here differ in their optical and HI properties. Two
are irregular galaxies, but most of them are known normal spirals;
NGC 2146 may be somewhat peculiar. Mrk348 is probably the galaxy
with the largest known extent of HI relative to optical size. It
was first noticed by Heckman, Balick & Sullivan (1978) in their
21-cm survey of Seyfert galaxies. Morris & Wannier (1980) obtained
a map of the HI with the Arecibo radio telescope and found it to
extend over an 8' x 12' region, which is an order of magnitude
larger than the Holmberg diameter of this nearly face-on galaxy
(Fig. 5). In the 21-cm observations at Westerbork (Heckman,
private communication) the gas shows an elongated distribution
suggestive of a bar with spiral arms. Its column density remains
fairly high ($\sim 5 \times 10^{19}$ cm^{-2}) out to the edge and is an order of

Table 1. Galaxies with very large HI disks

Galaxy	Type	Distance (Mpc)	D_{HI} (Kpc)	$\dfrac{D_{HI}}{D_{Ho}}$	M_{HI} ($10^9 M_\odot$)	$\dfrac{M_{HI}}{Lpg}$	Ref.
N5236 (M83)	Sc	8.9	>105	>3	19	0.17	1
N5457 (M101)	Sc	7.2	130	2.3	24	0.6	2
IC10	Ir	1.0	20	(6)	0.3	0.7	3
N4449	Ir	3.3	60	6	2	0.6	4
N628	Sc	10.6*	110	3	15	0.6	5
N2841	Sb	9.0*	>75	>2.5	2.8	0.16	6
N2146	Sabpec	14.0*	210	6	5.7	0.25	7
N262 (Mrk348)	Sa	62.0*	220	8	15	3	8

* Values based on Ho=75 km s^{-1} Mpc^{-1}
D_{HI} = Major HI diameter at 1×10^{19} cm^{-2}
D_{Ho} = Holmberg diameter (26.5 photographic mag arcsec^{-2})
M_{HI} = Total mass of neutral hydrogen
Lpg = Photographic luminosity
Ref.:1. Rogstad, Lockhart & Wright (1974), 2. Huchtmeier & Witzel
   (1979), 3. Huchtmeier (1979), 4. van Woerden, Bosma & Mebold
   (1975), 5. Briggs et al. (1980), 6. Bosma (1978), 7. Fisher
   & Tully (1976), 8. Morris & Wannier (1980).

magnitude higher than in the outer parts of M101. The rotation
velocity, 30/sin i km/s, gives a total mass of 1.3×10^{10} $M_\odot$/sin^{2}i
and an orbital period of 3×10^{10} sin i yrs in the outer parts.
This implies that, unless the system is very face-on (i<<30°),
most of its mass may be in neutral hydrogen. The gas in the outer
parts may not yet have completed one rotation, and may still be in
a primordial unrelaxed condition.

In conclusion, the present data suggest that disks generally
do not extend beyond 1 or 2 Holmberg radii, but there are excep-
tional cases where the HI is found at considerably larger distances. This gas may be primordial and under certain conditions fail
to fragment and to form stars (see e.g. Fall & Efstathiou, 1980;
cf. also end of section 4.4).

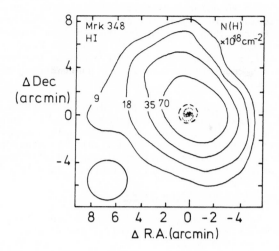

Fig. 5. HI map of Mrk348 (Morris & Wannier, 1980). The FWHP beam is shown by the circle. The optical image (dots) and the Holmberg size (dashed circle) are also indicated.

4.3 Kinematics

One of the major results of HI observations of galaxies has been the determination of the rotation curves at large radii, much beyond the visible disk. This has led to better estimates of the total mass and of the mass/luminosity ratio in the outer regions of galaxies. It has been often emphasized that rotation curves decline very slowly, if at all, even in the outermost parts (1 to 2 Holmberg radii) of the observed HI disk (Roberts, 1975b; Bosma, 1978). In fact, some drop-off is frequently seen to occur in the galaxies studied: it usually starts within the Holmberg radius, but is apparently not larger than 10 to 15% (~ 20 to 40 km/s) of the maximum velocity. There is also considerable variation among the various galaxies studied; in extreme cases such as NGC 891 (Sancisi & Allen, 1979) and NGC 4214 (Allsopp, 1979) the decrease might be larger and the curve may even become approximately Keplerian. The derived rotation velocities have uncertainties of the same order as the decrease (10 to 15% of the maximum). These uncertainties are partly due to observational effects (dependent on angular and velocity resolutions) and partly to large-scale deviations from axial symmetry ("oval" distortions and warps) or from central symmetry as in M81 or M101. The large-scale deviations from axial symmetry, such as shown in Fig. 4, often starting within the Holmberg radius, are generally interpreted as due to warping of the HI layer. "Tilted ring" models for the disk, as first proposed by Rogstad et al. (1974) for M83, with the assumption of circular motions, usually give a good representation of the observed deviations and lead to the derivation of flat rotation curves out to the last measured points. But the presence of non-circular motions in the plane or large-scale z-motions cannot be ruled out. Therefore the derived rotation curves, especially in the outer parts, remain quite uncertain.

159

The rotation curves are used together with photometric data to calculate the radial variation of the mass/luminosity ratio in galaxies. This seems to increase in the outer parts up to values of order 100 (cf. e.g. Bosma & van der Kruit, 1979) with an estimated uncertainty of a factor 3 due to the various uncertainties in the rotation curves, in the photometric data, and in the mass models.

In all known cases, including the systems with very extended HI listed in Table 1, the motion is well ordered, primarily with differential rotation, and the velocity dispersion is small, suggesting that the HI is in a flat disk rather than in "envelopes" or other peculiar configurations. The distribution in z, i.e. perpendicular to the equatorial plane, is discussed below.

4.4 Thickness of gas disks

In our Galaxy, the thickness of the HI layer, Δz (between half-maximum-density points), rises gradually with radius R from $\Delta z \sim 100$–200 pc at $R < 4$ kpc, to $\Delta z \sim 250$ pc at $4.5 < R < 10$ kpc, and has a steep increase beyond $R \sim 12$ kpc where $\Delta z > 1$ kpc (Henderson, 1979) (see Fig. 6). In other galaxies little is known yet; but in the edge-on systems which have been studied with sufficient angular resolution, NGC 891 (Sancisi & Allen, 1979), and IC 2233 (Sancisi & de Vaucouleurs, in preparation) there is evidence for a similar increase of Δz from a few hundred parsecs in the inner parts to more than 1 kpc in the outer parts. This probably occurs also in M31, as claimed by Whitehurst, Roberts & Cram (1978) and discussed by Baldwin in a chapter of this volume.

This has interesting implications for the three-dimensional distributions of total mass and the HI volume density in the outer parts of disk galaxies. Detailed investigations of M33 (Newton, 1980a), M31 (Newton & Emerson, 1977), IC 342 (Newton, 1980b) and of the almost completely face-on galaxy NGC 3938 (van der Kruit, Shostak & van Albada, 1979) have indicated that the velocity dispersion remains constant at about 10 ± 2 km/s over the whole disk, including its outer parts. From this it follows (cf. Kellman, 1972) that the outer thickening of the gas layer, observed in edge-on galaxies, must be related to a radial decrease of the total mass density in the plane. In NGC 3938 this decrease is found to be faster than would follow from its flat rotation curve. In an edge-on system like NGC 891, where both the rotation curve and the increase of thickness of the HI layer are measured, it should be possible to put constraints on the three-dimensional distribution of mass.

The slow decrease of HI column density in the outer parts of the disk, already mentioned above, may lead to the incorrect conclusion that the average volume densities also decrease slowly. The outer thickening of the HI layer, by as much as factors of 5

or 10, which may occur rather suddenly near the edge of the bright
optical disk, would imply a correspondingly drastic drop in the
average volume density of the gas. This may be one of the causes
for the lack of star formation in the outer parts of galaxies
and, consequently, for their sharp edges (cf. van der Kruit &
Searle, 1980).

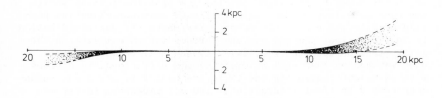

Fig. 6. Idealized picture of a section of the HI layer in our
Galaxy, showing its outer warping and thickening (cf. Henderson,
1979).

4.5 Warps

In several galaxies the hydrogen layer shows a significant
departure from the flat principal plane, generally outside the
bright optical disk. This was first found in the early 21-cm surveys of our Galaxy (see Fig. 6). It is also seen in the other two
spirals of the Local Group, M31 and 33, and in almost all edge-on
galaxies studied to date (NGC 4244, 4565, 4631, 5907; possible
exceptions are NGC 891, 3556 and 7640). It is also inferred for
many, less inclined galaxies (M83, NGC 300, 2841, 5033, 5055,
6503, 7331 and IC 342) from their non-axisymmetric HI distributions and kinematics (for a recent review see van Woerden, 1979).
Besides the integral-sign morphology, very little is known yet
about the structure and extent of the warping. The statistics
suggest that warps are not induced by major nearby companions, but
have some other cause and must be able to survive for several
rotation periods.

The proposed explanations for galactic warps have involved:
flow through an intergalactic medium (Kahn & Woltjer, 1959), free
precession of the stellar disk (Lynden-Bell, 1965) tidal interactions (Hunter & Toomre, 1969) and recent infall of HI gas (Rogstad, Wright & Lockhart, 1976). In the more recent work halos of
various sizes, masses and shapes have been invoked to explain the
origin and/or persistence of warps (Binney, 1978; Saar, 1979;

R. SANCISI

Tubbs & Sanders, 1979; Bertin & Mark, 1980).

4.6 Asymmetries

Large-scale deviations from axial symmetry often occur in disk galaxies, as shown both in the kinematics and in the distribution of the gas (cf. e.g. Bosma, 1978). These may be due to so-called oval distortions and/or warping of the disk. But there are also large-scale asymmetries with respect to the galaxy centre, which are often shown by the light and the gas distributions as well, and even sometimes by the radio continuum emission. They are particularly noticeable in the outer parts where the gas distribution is lopsided, i.e. the gas extends farther out on one side of the galaxy than on the other. Baldwin, Lynden-Bell & Sancisi (1980) have emphasized that such asymmetries are often found in systems with no bright companions and therefore cannot be transient phenomena caused by tidal interactions. The lop-sidedness is rather pronounced in M101, IC 342, NGC 628, 891, 2841 and 6503, and milder in M83, NGC 5033, 7331, 3198, and 4565. Kinematical models such as those suggested by Baldwin et al. (1980) may explain, at least to some extent, the long persistence of the asymmetry against "winding up" by differential rotation. The lop-sidedness of the gas may, however, hint at an asymmetric distribution of the total mass of the system (disk and halo) with respect to its centre.

A lop-sidedness, such as that observed in nearby galaxies, may also exist in our own Galaxy and it could account, perhaps, for some of the yet unexplained aspects of 21-cm studies (cf. Kerr, 1969; Burton, 1974). These include the difference between northern and southern rotation curves, the systematic difference in the cut-off velocities, and the presence of anomalous, non-circular velocities (e.g. at $\ell=90°$).

5. INTERGALACTIC HI

The presence of truly isolated clouds of neutral hydrogen of galactic mass in intergalactic space has never been established unambiguously, despite many attempts made in recent years with the largest radio telescopes and the most sensitive receivers. The only objects thus far found which come close to such a population are (i) some hydrogen-rich dwarf galaxies of extremely low luminosity and, (ii) gas complexes in the neighbourhood of interacting systems. The latter, however, are likely to be the result of recent tidal stripping.

5.1 Searches for intergalactic HI; the hydrogen-rich dwarfs

Most published searches for isolated HI clouds located far from optically visible galaxies have been made with the 91-m telescope of the National Radio Astronomy Observatory (Roberts & Steigerwald, 1977; Shostak, 1977; Fisher & Tully, private communication). These searches failed to detect unseen galaxies of masses in the range of 10^7 to 10^{11} $M_\odot$ for distances between 1 and 40 Mpc respectively. The 21-cm surveys of some nearby groups of galaxies (Lo & Sargent, 1979; Materne, Huchtmeier & Hulsbosch, 1979; Haynes & Roberts, 1979) give upper limits of about 0.5 to 1×10^8 $M_\odot$ for narrow velocity features ($\Delta V \approx 35$ km/s) at distances less than 10 Mpc. But in some selected areas, Lo and Sargent were able to reach levels an order of magnitude lower using the Effelsberg 100-m telescope and discovered four objects that turned out to be uncatalogued, intrinsically faint dwarf irregular galaxies, which are barely detectable on the Palomar Sky Survey prints. The most extreme of these is the one designated M81 dwA. This is the smallest dwarf irregular galaxy yet detected and may be composed mostly of neutral hydrogen. It has Mpg ~ -9, an optical size of 1 Kpc and a total HI mass of 1×10^7 $M_\odot$. Its M_H/L_{pg} ~ 20 is the highest value of any galaxy. The neutral hydrogen is distributed in a lumpy ring of 2 kpc diameter around the optical object (Lo, Sancisi & Sargent, in preparation). The rotation is only 3.5 ± 1 km/s and the velocity dispersion is 6 ± 1 km/s.

The findings of Lo and Sargent may imply that a large population of such dwarf galaxies with large M_H/L ratios, and therefore properties approaching those of intergalactic HI clouds, exist near galaxies or in galaxy groups. Their study is being followed up at present with photographic surveys and 21-cm line observations. These objects presumably represent the extreme, faint end of the luminosity and mass functions of normal galaxies. Their properties and space density are also interesting in light of the effect that their capture may have on the evolution of galaxies.

5.2 Gas near interacting systems

There are several known cases of HI "streams" or "cloud complexes" in the neighbourhood of interacting galaxies (Table 2). The objects in this list and the associated gas clouds show a wide range of properties as indicated in Table 3. The majority of these clouds are not completely resolved either in space or velocity; therefore the actual densities in knots may reach considerably larger values than given in the table, and the velocity dispersions are probably smaller. Usually these clouds contain only a fraction (~ 10 to 20%) of the gas in the galaxies; Stephan's Quintet, with a total HI mass of ~ 10^{10} $M_\odot$, represents an extreme case in which nearly all the gas is found outside the galaxies. Quite narrow velocity profiles, as found in the long "plume" of NGC 3628 ($\Delta V < 17$ km/s), are not uncommon. There is almost always a continuity in space and velocity between the HI complexes and the disks of the galaxies as in the case of NGC 4631/56/27, illustrated in Fig. 7.

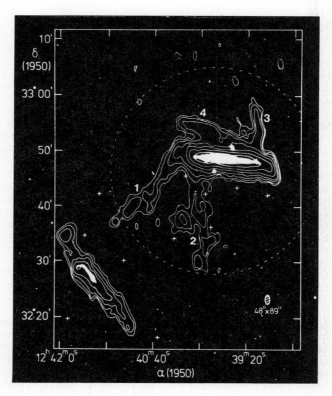

Fig. 7. HI map of the interacting system N4631/4656/4627
(Weliachew et al., 1978). The contour values are 0.5, 1, 2, 4, 8,
16, 32, 64×10^{20} atoms cm^{-2}.

In a few cases (e.g. the Antennae and Leo Triplet) the HI even has
a faint optical counterpart extending away from the galaxy. In
other instances the connection with the galaxies is not so
obvious, and the doubt remains that they may be unrelated, intergalactic clouds or even foreground emission associated with our
Galaxy. The latter possibility seems to be presently favoured for
the clouds in the Sculptor Group (Haynes & Roberts, 1979). For the
HI cloud near NGC 1023 discovered by Hart et al. (1980) at Jodrell
Bank, the preferred interpretation is that of a background
isolated object. However, a recent, more detailed study (Sancisi,
et al., in preparation) has revealed velocity and spatial
continuity with the gas around NGC 1023 (Fig. 1) suggesting a
connection with the galaxy and a tidal origin. This is also supported by the presence near the galaxy of some patchy optical
emission and of a dwarf companion. The latter was noted by Arp and
led him to include the NGC 1023 system in his Atlas of Peculiar
Galaxies (see also discussion on S0 galaxies above).

Table 2. Interacting systems with intergalactic HI

Galaxy and Magellanic Clouds	Mathewson, Cleary & Murray, 1974
M81/M82/N3077	Van der Hulst, 1979b
N4631/N4656/N4627	Weliachew, Sancisi & Guèlin, 1978
N3623/N3627/N3628 (Leo Triplet)	Haynes, Giovanelli & Roberts, 1979
N4038/N4039 (Antennae)	Van der Hulst, 1979a
N55/N300 (Sculptor group)	Mathewson, Cleary & Murray, 1975
N5194/N5195 (M51)	Haynes, Giovanelli & Burkhead, 1978
Stephan's Quintet	Allen & Sullivan, 1980
N1023	Hart, Davies & Johnson, 1980
N3226/N3227	Knapp, Kerr & Williams, 1978
N3395/N3396	Shostak & Rose, private communication

Table 3. Properties of HI clouds near interacting systems

Masses, M_H	$1\text{-}100 \times 10^8$	$M_\odot$
Column densities, N_H	$1\text{-}50 \times 10^{19}$	cm^{-2}
Volume densities, n_H	$0.01\text{-}1$	cm^{-3}
Linear sizes	$10\text{-}100$	kpc
Velocity widths, ΔV	$15\text{-}50$	km/s

In conclusion, it appears that almost all galaxies giving the impression of strong tidal interactions, such as those depicted in Arp's Atlas, are found to be accompanied by intergalactic HI clouds when observed with sufficient sensitivity. This has led to the hypothesis, supported by computer simulations, that this gas has been drawn far out from the disks of the galaxies involved by the tidal interactions. An alternative explanation is that most of this intergalactic HI may be primordial gas left over from galaxy formation.

ACKNOWLEDGMENTS

I am grateful to M. Fall and G.S. Shostak for helpful comments on the manuscript, to T.S. van Albada, J.S. Gallagher, P.C. van der Kruit, R.H. Sanders, U.J. Schwarz and H. van Woerden for valuable discussions.

REFERENCES

Allen, R.J. & Goss, W.M. (1979). Astron. Astrophys. Suppl. 36, 135.
Allen, R.J. & Sullivan, W.T. III (1980). Astron. Astrophys. 84, 181.

R. SANCISI

Allsopp, N.J. (1979). Mon. Not. R. Astr. Soc. 188, 765.
Baldwin, J.E., Lynden-Bell, D. & Sancisi, R. (1980). Mon. Not. R. Astr. Soc. in press.
Bertin, G. & Mark, J.W.-K. (1980). Astron. Astrophys. 88, 289.
Binney, J. (1978). Mon. Not. R. Astr. Soc. 183, 779.
Bosma, A. (1978). Ph.D. Thesis, Groningen University.
Bosma, A., Goss, W.M. & Allen, R.J. (1980). Astron. Astrophys. in press.
Bosma, A., Hulst, J.M. van der & Sullivan, W.T. III (1977). Astron. Astrophys. 57, 373.
Bosma, A. & Kruit, P.C. van der (1979). Astron. Astrophys. 79, 281.
Briggs, F.H., Wolfe, A.M., Krumm, N. & Salpeter, E.E. (1980). Astrophys. J. 238, 510.
Burton, W.B. (1974). In Galactic and Extra-Galactic Radio Astronomy p.82, Eds. G.L. Verschuur & K.I. Kellermann, Springer-Verlag, New York.
Burton, W.B. (1976). Ann. Rev. Astron. Astrophys. 14, 277.
Faber, S.M. & Gallagher, J.S. (1979). Ann. Rev. Astron. Astrophys. 17, 135.
Fall, S.M. & Efstathiou, G. (1980). Mon. Not. R. Astr. Soc. 193, 189.
Fisher, J.R. & Tully, R.B. (1976). Astron. Astrophys. 53, 397.
Gouguenheim, L. (1979). In Photometry, Kinematics and Dynamics of Galaxies, p.201. Ed. D.S. Evans, University of Texas at Austin.
Hart, L., Davies, R.D. & Johnson, S.C. (1980). Mon. Not. R. Astr. Soc. 191, 269.
Haynes, M.P., Giovanelli, R. & Burkhead, M.S. (1978). Astron. J. 83, 938.
Haynes, M.P., Giovanelli, R. & Roberts, M.S. (1979). Astrophys. J. 229, 83.
Haynes, M.P. & Roberts, M.S. (1979). Astrophys. J. 227, 767.
Heckman, T.M., Balick, B. & Sullivan, W.T. III. (1978). Astrophys. J. 224, 745.
Henderson, A.P. (1979). IAU Symp. 84. The Large-Scale Characteristics of the Galaxy, p. 493. Ed. W.B. Burton, Reidel Dordrecht Holland.
Holmberg, E. (1958). Medd. Fr. Lunds Astr. Obs. Ser. II., No. 136.
Huchtmeier, W.K. (1979). Astron. Astrophys. 75, 170.
Huchtmeier, W.K. & Witzel, A. (1979). Astron. Astrophys. 74, 138.
Hulst, J.M. van der (1979a). Astron. Astrophys. 71, 131.
Hulst, J.M. van der (1979b). Astron. Astrophys. 75, 97.
Hunter, C. & Toomre, A. (1969). Astrophys. J. 155, 747.
Huntley, J.M. (1978). Astrophys. J. 225, L101.
Kahn, F.D. & Woltjer, L. (1959). Astrophys. J. 130, 705.
Kellman, S.A. (1972). Astrophys. J. 175, 353.
Kerr, F.J. (1969). Ann. Rev. Astron. Astrophys. 7, 39.
Knapp, G.R., Kerr, F.J. & Williams, B.A. (1978). Astrophys. J. 222, 800.
Kotanyi, C.G. & Ekers, R.D. (1979), Astron. Astrophys. 73, L1.

Kruit, P.C. van der & Allen, R.J. (1978). Ann. Rev. Astron. Astrophys. 16, 103.

Kruit, P.C. van der & Searle, L. (1980). Astron. Astrophys. in press.

Kruit, P.C. van der, Shostak, G.S. & Albada, T.S. van (1979). In Photometry, Kinematics and Dynamics of Galaxies, p.277. Ed. D.S. Evans, University of Texas at Austin.

Lo, K.Y. & Sargent, W.L.W. (1979). Astrophys. J. 227, 756.

Lynden-Bell, D. (1965). Mon. Not. R. Astr. Soc. 129, 299.

Materne, J., Huchtmeier, W.K. & Hulsbosch, A.N.M. (1979). Mon. Not. R. Astr. Soc. 186, 563.

Mathewson, D.S., Cleary, M.N. & Murray, J.D. (1974). Astrophys. J. 190, 291.

Mathewson, D.S., Cleary, M.N. & Murray, J.D. (1975). Astrophys. J. 195, L97.

Morris, M. & Wannier, P.G. (1980). Astrophys. J. 238, L7.

Newton, K. (1980a). Mon. Not. R. Astr. Soc. 190, 689.

Newton, K. (1980b). Mon. Not. R. Astr. Soc. 191, 169.

Newton, K. & Emerson, D.T. (1977). Mon. Not. R. Astr. Soc. 181, 573.

Raimond, E., Faber, S.M., Gallagher, J.S. III & Knapp, G.R. (1980). S.R.Z.M. Preprint.

Roberts, M.S. (1975a). In Stars and Stellar Systems Vol. IX, p.309. Eds. A. Sandage, M. Sandage & J. Kristian, The University of Chicago Press, Chicago.

Roberts, M.S. (1975b). IAU Symp. No. 69. Dynamics of Stellar Systems, p.331. Ed. A. Hayli, Reidel Dordrecht Holland.

Roberts, M.S. & Steigerwald, D.G. (1977). Astrophys. J. 217, 883.

Rogstad, D.H., Lockhart, I.A. & Wright, M.C.H. (1974). Astrophys. J. 193, 309.

Rogstad, D.H., Wright, M.C.H. & Lockhart, I.A. (1976). Astrophys. J. 204, 703.

Rose, W.K. & Tinsley, B.M. (1974). Astrophys. J. 190, 243.

Rots, A.H. (1975). Astron. Astrophys. 45, 43.

Saar, E.M. (1979). IAU Symp. 84. The Large-Scale Characteristics of the Galaxy, p. 513. Ed. W.B. Burton, Reidel Dordrecht Holland.

Sancisi, R. (1978). IAU Symp. 77. Structure and Properties of Nearby Galaxies, p.27. Eds. E.M. Berkhuysen & R. Wielebinski, Reidel Dordrecht Holland.

Sancisi, R. & Allen, R.J. (1979). Astron. Astrophys. 74, 73.

Sancisi, R., Allen, R.J. & Sullivan, W.T. III. (1979). Astron. Astrophys. 78, 217.

Sandage, A. (1961). The Hubble Atlas of Galaxies, Carnegie Institution of Washington, Washington D.C.

Sanders, R.H. (1979). IAU Symp. 84. The Large-Scale Characteristics of the Galaxy, p. 383. Ed. W.B. Burton, Reidel Dordrecht Holland.

Sanders, R.H. (1980). Astrophys. J. in press.

Schechter, P.L. & Gunn, J.E. (1979). Astrophys. J. 229, 472.

Shane, W.W. (1980). Astron. Astrophys. 82, 314.

Shostak, G.S. (1977). Astron. Astrophys. 54, 919.

Shostak, G.S., Woerden, H. van & Schwarz, U.J. (1979). In Photometry, Kinematics and Dynamics of Galaxies, p.213. Ed. D.S. Evans, University of Texas at Austin.

Tubbs, A.D. & Sanders, R.H. (1979). Astrophys. J. 230, 736.

Unwin, S.C. (1980). Mon. Not. R. Astr. Soc. 190, 551.

Weliachew, L. Sancisi, R. & Guélin, M. (1978). Astron. Astrophys. 65, 37.

Whitehurst, R.N. & Roberts, M.S. & Cram, T.R. (1978). IAU Symp. 77. Structure and Properties of Nearby Galaxies, p.175. Eds. E.M. Berkhuysen & R. Wielebinski, Reidel Dordrecht Holland.

Woerden, H. van (1977). In Topics in Interstellar Matter, p.261. Ed. H. van Woerden, Reidel Dordrecht Holland.

Woerden, H. van (1979). IAU Symp. 84. The Large-Scale Characteristics of the Galaxy, p.501. Ed. W.B. Burton, Reidel Dordrecht Holland.

Woerden, H. van, Bosma, A. & Mebold, U. (1975). In La Dynamique des Galaxies Spirales, p.483. Colloque du C.N.R.S. No. 241, Ed. L. Weliachew.

RADIO CONTINUUM EMISSION FROM GALAXIES

R.D.Ekers

Kapteyn Astronomical Institute, University of Groningen,
Groningen, The Netherlands

1 INTRODUCTION

In this chapter I intend to concentrate on the highlights of
the last few years' work on the structure of the continuum emission from galaxies and to refer to other reviews in areas where
there is relatively little new. Ekers (1975) discussed the radio
emission from spiral and elliptical galaxies at a time when the
first maps of the continuum emission from spiral galaxies were
being made. Further reviews of the distribution of emission in
spiral galaxies are given by van der Kruit & Allen (1976) and van
der Kruit (1978). A recent discussion of the distribution in z
is given in Sancisi and van der Kruit (1980).

Reviews of the nuclei of normal galaxies are given by Ekers
(1978a, 1979) and van der Hulst (1979), with the latter including
the first results from the higher resolution VLA observations.

For the elliptical galaxies it is now becoming clear that
the "normal" radio emission is the well collimated ejection of
radiating plasma. Since this phenomenon is presumably related to
other basic properties of the elliptical galaxies, e.g. the dynamics, nuclei, gas content or environment, I have included a discussion of some of these topics in Section 5.

One of the most obvious things we have learned from the
external galaxies is that their radio properties are very diverse
– there is a range of ~100 in average disk brightness and even
more in the central few kpc. Hence an individual galaxy cannot
tell us much about the amount of non-thermal emission; much
clearer results are coming from the statistical studies discussed
in Section 4.2.

1.1 Surveys

Many of the new results discussed have come from the excellent radio surveys which have recently been completed. Their

properties are summarised in Table 1. Noteworthy among these
surveys are the large sample of pairs in the survey of Stocke et
al. (1978) which gave the first strong indication of enhanced
emission from multiple systems (see 4.2), the very large and sensitive survey of galaxies in the Arecibo declination range (Dressel
and Condon 1978) and the recent Westerbork surveys by Hummel
(1980) and Kotanyi (1980) which have sufficient angular resolution
to separate disk and central components for the whole sample.

TABLE 1. Normal galaxy surveys since 1977

					Radio observations			
Optical selection					Freq.		Flux limit	
Cat.	Type	m_p	Dec.	No.	(GHz)	θ	(mJy)*	Ref.
RCBG	S,Ir	<12	>−20°	181	2.7	10"	20	1
RCBG	E,S0	−	<0°	181	5.0	4'.0	12	2
Kar.	pairs	<15.7	>−3°	603	2.7	2'.5	40	3
UGC	all	<14.5	0,37°	2095	2.4	2'.7	15	4
UGC	all	<14.5	0,37°	168	2.7	6"	35	5
RCBG	all	<12	−30,30°	400	1.4	20"	10 point	6
"	"	<13	>30°				50 disk	
UGC	S	<12	20,60°	91	0.4	3×5'	70	7
HMS	E,S0	<−20+	−18,60°	43	0.4	3×5'	60	8
RCBG	all	<14	Virgo	274	1.4	23"	~10	9

* 1 mJy = 10^{-29} Wm^{-2} Hz^{-1}
+ M_{pg}

Ref.: 1 Crane (1977), 2 Disney & Wall (1977), 3 Stocke et al.
(1978) 4 Dressel & Condon (1978), 5 Condon & Dressel (1978)
6 Hummel (1980a,d), 7 Gioia & Gregorini (1980), 8 Feretti &
Giovannini (1980), 9 Kotanyi (1980).

1.2 Emission mechanism

The continuum radio emission detected from external galaxies
is produced either by thermal processes or by the synchrotron
process with the latter dominating in most cases. For the synchrotron process the volume emissivity is given by

$$\epsilon(\nu) = n\ B^{\frac{1+\gamma}{2}}\ \nu^{\frac{1-\gamma}{2}} = n\ B^{1-\alpha}\ \nu^{\alpha}$$

where the number of relativistic electrons with energy E is
$N(E) = n\ E^{-\gamma}$, B is the magnetic field and ν the frequency.
Hence an observation of the radio continuum is a measurement of
the product of the magnetic field strength and relativistic
particle density. The density and distribution of relativistic
particles depends on the sources, the losses and the rate at
which they can diffuse through or out of the galaxy. The role
played by the magnetic field in normal galaxies is largely
unknown, its effects are often ignored even though it may be the
dominant force. The sources of relativistic particles and their
possible reacceleration is speculative and the diffusion parameters are in dispute.

Consequently, the interpretation of observations of the
radio continuum emission from galaxies can not be straight
forward, yet this information may be still very valuable. For
the spiral galaxies it can give information on conditions in the
interstellar gas, such as regions of compression or acceleration.
It may indicate the evolutionary history of the galaxy, especially if supernovae are significant sources of particles or magnetic
field. It is also an indicator of more violent activity such as
in the nuclei of galaxies and in the radio galaxies.

2   DISKS OF SPIRAL GALAXIES

2.1 The Galaxy

To discuss our Galaxy in the context of external galaxies we
are mostly interested in the large scale properties. The radio
emission from discrete sources e.g. stars, planets, pulsars, etc.
has a negligible integrated effect. Supernova remnants also make
a negligible contribution to the total flux but some are strong
enough to be individually detectable in external galaxies. More
importantly, they may play an indirect role as a possible source
of relativistic electrons and also perhaps magnetic fields.
Thermal emission is seen both from discrete HII regions and from
the gaseous disk but even the biggest galactic HII regions would
be detectable only in external galaxies closer than a few Mpc.

The dominant radio emission is the diffuse non-thermal
component from the whole disk of the Galaxy. In principle this
can be separated from the thermal emission on the basis of
spectra and polarization but there is still considerable
uncertainty in this separation. The non-thermal emission is
almost certainly due to the synchrotron process as the cosmic ray
electrons interact with an interstellar magnetic field of order

10^{-6} gauss. It is clear from the large scale angular distribution of this emission that most of it originates in the disk of the Galaxy and, with less certainty, it is also thought that this emission is strongest in the spiral arms.

2.2 Disk emission from spiral galaxies

Although the observation of the non-thermal emission in our Galaxy is important for comparison with cosmic-ray and gamma-ray data the greatest advances in our understanding are coming from the good angular resolution observations of external galaxies. These observations are not hampered by looking from the inside and allow us to investigate the properties for a sample of galaxies of different types.

The spectacular spiral structure in the M51 radio continuum (Mathewson, van der Kruit & Brouw 1972) shows that some of the non-thermal emission is associated with phenomena related to the formation of the spiral structure - eg compression of the inter-galactic medium and its magnetic field, acceleration in the shocks, or an additional source of relativistic particles. Surprisingly, no further clear cases of this special enhanced condition have been seen, and Hummel (1980) shows that the presence of spiral arms does not make an important contribution to the total non-thermal emission (Sect. 4.2).

The radial distribution of continuum radio emission drops off much faster than the distribution of gas (see especially Fig. 1 in Sancisi & Allen 1979). The exact radial dependence of the non-thermal component is more difficult to establish since it is necessary to correct for the thermal contribution. Van der Kruit et al (1977) and van der Kruit (1977) claim a good correlation between the non-thermal emission and the distribution of optical brightness indicating a correlation with the total stellar disk population rather than the extreme population I. This suggests that either the sources of relativistic particles are distributed like the old disk population or that the old disk population effects the distribution of volume emissivity indirectly. E.g. via the mass distribution which may control the particle containment.

New results on the polarization of the non-thermal emission (Segalovitz, Shane & de Bruyn 1976 and Beck, Berkhuysen & Wielebinski 1980) indicate a large scale regularity in the magnetic field structure and again this may be reminding us of the importance of the magnetic fields in spiral galaxies.

2.3 The distribution of emission in z - the radio haloes

To study the distribution of radio emission perpendicular to the plane of the galaxy we need to observe an edge-on system.

This can be done in our Galaxy e.g. Ilovaisky & Lequeux (1972) suggest two components with a thickness of 500 pc and 2 Kpc but the results are somewhat model dependent (Baldwin 1976). A recent evaluation of the z distribution in other galaxies is given by Sancisi & van der Kruit (1980). They conclude that although thick disks (6 - 8 kpc) exist with steepening spectral index further above the plane, the observations are not yet sufficient to establish the generality of the phenomenon or to enable detailed comparisons with the theoretical models.

3 CENTRES OF GALAXIES

3.1 Sagitarius A

A clear component of the Galactic continuum radiation is the complex distribution seen in the direction of the Galactic centre. This emission has three main components: an extended source about 180 x 70 pc elongated along the Galactic plane which is probably non-thermal, a complex of giant HII regions in a similar area and another non-thermal source, Sgr A, about 9 x 7 pc very close to the centre of the Galaxy. Higher resolution observations (Ekers et al. 1975) show that Sgr A is also complex, consisting of two main features: Sgr A East which has a steep non-thermal spectrum and Sgr A West which is a flatter spectrum centrally peaked source ~3 pc in size. It is very similar to the distribution of infra-red emission from dust in the 10 - 50 μ range and it is probably thermal emission coming from gas surrounding the luminous stars.

Near the peak of SgrA West is a much smaller source (Balick & Brown 1974) with diameter <0".001 (2 × 10^{14} cm) (Geldzahler, Kellerman & Shaffer 1979). The nature of this point source is still unclear. Assuming that the measured angular size is the intrinsic size Brown et al.(1978) present arguments against a synchrotron model with self absorption and suggest instead an analogy with radio binaries. Others have commented on the similarity with the Ryle et al.(1978) class of compact sources (including SS433). It is also reasonable to draw an analogy with the compact sources seen in other galaxies (based on its location, spectrum, size and lack of strong variability) but then we have to consider models also capable of supplying millions of times more energy.

3.2 Structure in the central region

Although we can always obtain the most detailed information from observations of our Galactic centre we cannot tell whether these features are peculiar to us without looking at other galaxies. In most cases the linear resolution available for a study of external galaxies is not better than a few thousand

parsecs so I will use the term "central sources" when referring
to structures on this scale, reserving the term "nucleus" for
those sources known to be less than a few parsec in size.

Recent observations, especially with the VLA, show that
these central sources are complex regions of emission similar to
the inner region of our Galaxy. Van der Hulst (1979) gives a
distribution of linear sizes which also indicates that the
Galactic centre is quite typical.

3.3 Nuclear radio sources

The increasing dominance of the central sources with earlier
type is accompanied by a flattening of the average spectral index
(Ekers 1978b but NB incorrect sign of the spectral index in the
figure) and an increasing number of cases in which the central
source is known to contain a very compact component <0.1 pc in
size (Kellermann et al 1976, Crane 1979, van Breugel et al 1980).
Most of these nuclear radio sources are found in E and S0
galaxies. In a few cases they show some variability on time
scales of a few years (NGC4552 - Sramek 1975, NGC1052 and NGC5077
- Ekers and Heeschen private communication) but most have
remained constant for the last 10 years. Weak variability has
been suggested for two of the spiral galaxies with nuclear
sources, M81 and M104 (de Bruyn et al. 1976).

4   STATISTICAL PROPERTIES OF GALAXIES

4.1 Method of analysis

It is necessary to remove the effects of observational
selection before it is possible to give a useful analysis of the
statistical data from the surveys in Table I. Fortunately most
of these surveys have well defined limits and for many of the
galaxies the distance is also known so it is possible to remove
the main selection effects. In the procedure used in Hummel
(1980b) the fractional radio luminosity function is derived for
the sub classes of galaxies to be compared. The important aspect
of this procedure is that in addition to calculating the radio
power the survey limits are used to calculate also the range of
observable power (cf V/V_{max} Schmidt 1968). The fraction of
galaxies in the sample which have this power is then determined
by comparing the number of detections with the number of galaxies
in the sample which could have been detected at this level.

4.2 Correlations

Although individual galaxies show a large range in the ratio
of disk to central source radio emission, on average the central
source contains only 10% of the total flux (Hummel 1980a,b).

174

Consequently, statistical inferences from surveys which measure
the total flux from a galaxy apply to the disk emission.

The following summary of the correlation between radio emission and other properties is based on the main conclusions from
the survey by Hummel (1980a,b). In this analysis the disks and
central sources are separated.

TABLE 2. Correlations with radio emission.

Parameter	Disk		Centre	
Optical luminosity	Yes	$\langle L_r \rangle \propto \langle L_o \rangle$	Yes	$\langle L_r \rangle \propto \langle L_o \rangle$
Hubble type	Yes	(max at Sbc)	Yes	(max at Sa)
Bar	No		Yes	
Colour	No	(Sb to Scd)	–	
Luminosity class (arm development)	No		–	
Companions	No		Yes	
Disk brightness	–		No	

The dominant correlation is that between radio power and
optical luminosity (Cameron 1971, Crane 1977, Hummel 1980a,b) and
this must be taken into account when looking for any other
effects. Since the correlation is linear this can be done conveniently by using the ratio of radio to optical luminosity. All
the additional tests in Table 2 refer to this ratio.

The disk and the central source have different dependences on
Hubble type. This is clearly seen in Fig. 1 in which the data
from Hummel (1980b) is combined with that from Hummel and Kotanyi
(in preparation) for the SO galaxies. The central sources have
increasing dominance towards early Hubble types. Even though the
average ratio of radio to optical luminosity is low for the central sources in SO galaxies they are still the dominant component
since no disk emission has been detected in any SO galaxy.

The lower disk emission in the very early types is not too
surprising since they have much less gas and consequently less
ability to confine the relativistic electrons and magnetic
field.

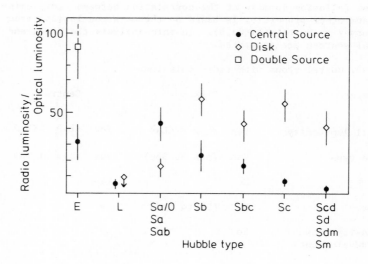

Fig. 1. The 10 percentile of the radio luminosity/optical luminosity distributions for various Hubble types (i.e. 10% of the galaxies of a given type have a ratio larger than the value plotted).

An important negative result is the lack of correlation between radio emission and either colour or spiral arm development. This shows that the young population is not making a significant contribution to the non-thermal emission (cf Section 2.2)

After a number of conflicting results Stocke (1978) clearly showed that galaxies with companions had higher radio emission and Condon & Dressel (1978) noted that the enhanced emission came from the nucleus in some cases. These results were confirmed by Hummel(1980c) who also shows that the effect is almost entirely due to the central sources. The enhanced central emission in both interacting and barred galaxies may be triggered by the additional infalling gas. A small number of strongly interacting systems also have peculiar radio structure (Allen, Ekers, Burke & Miley 1973, Stocke et al. 1978), which may be a direct result of the interaction but these are too rare to contribute to the statistical result given above.

4.3 Spectra

Good quality measurements of the flux densities of galaxies are now being made at Bonn (Klein & Emerson, private communication) and NRAO (van der Hulst, 1979). These are showing that most galactic disks have similar power law spectra with a mean spectral index of −0.7 and relatively little dispersion. At 1.4 GHz <13% of the average radio flux can be thermal. The distribution of the radio spectral indices of the disks of galaxies now looks surprisingly similar to that seen in the lobes of extra galactic radio sources, suggesting that a common mechanism may be determining the energy distribution of relativistic particles (reacceleration ?)

5 ELLIPTICAL GALAXIES

5.1 Total emission

If we look at the radio properties of the nearby ellipticals we find a situation considerably different from that for the spiral galaxies. For example NGC5128 (Cen A), the nearest giant elliptical galaxy, is a thousand times more powerful a radio source than the brightest spiral galaxies and furthermore its radio emission comes from a multiple lobed radio structure which bears no resemblance to the optical light distribution. In most of the literature a clear distinction is made between these "radio galaxies" (mostly found in radio surveys) and "normal galaxies" (mostly found in galaxy surveys) but this is mainly due to the shapes of the radio luminosity function. A question which then arises is whether at lower levels we can detect radio emission coming from the optical image of the elliptical galaxies and which may be more closely related to the kind of emission seen in the spiral galaxies.

Ekers and Kotanyi (1978) showed that one elliptical galaxy, NGC4472 which is just as weak as a spiral, still has the double lobe structure. Similar results are now found in other low luminosity ellipticals: NGC3665 (Kotanyi 1979), NGC1052 (Heeschen & Wrobel, private communication), NGC1399 (Bosma, private communication).

5.2 Elliptical / SO problem

Since there is a dramatic change in the properties of the radio continuum emission going from spiral to elliptical galaxies (Ekers 1975) an obvious question to ask is: what kind of radio sources are the transition class, the SO galaxies? This has turned out to be a difficult question to answer because of the elusive nature of the SO classification. The distribution of apparent flattening for the detected SO galaxies from Dressel &

Condon (1978) is quite different from that of SOs in general, in the sense that many more detected SO galaxies were face on. Since it is unlikely that the radio emission is highly anisotropic it seems more likely that the detected "face on" SOs are mostly elliptical galaxies. Furthermore, all the SO galaxies which have been found to have extended radio emission have the double lobed structure typical of the elliptical galaxies. On closer investigation it has been found that these are often elliptical galaxies which have been classified as SO because they also show a band of dust (eg NGC3665, Centaurus A, Fornax A).

5.3 Radio source ejection and dynamics

Since the normal state of radio emission from an elliptical galaxy is a well collimated structure it defines an additional axis which may help in our understanding of elliptical galaxies.

Various indirect arguments (e.g. Miley 1980) suggest that the radio axis should also be the rotation axis of the galaxy and this is supported by observations of the rotation axes of radio galaxies (e.g. Simkin 1979) although some non aligned cases are also found (e.g. Jenkins & Scheuer 1980). It has also been found that a number of the radio emitting elliptical galaxies contain dust bands indicating the presence of a gaseous disk which is perpendicular to the radio axis in a majority of cases (Kotanyi & Ekers 1979). Some of these radio sources have a twisted (S shape) structure perhaps indicating a rotation of the axis of ejection during the life of the radio source. It can be conjectured that this rotation has occurred as a result of material being captured by an elliptical galaxy whose principal axes are not the same as the rotation axis of the captured material. This will cause the captured material to precess in the gravitational field of the elliptical galaxy until dissipation causes it to line up with a principal axis of the elliptical galaxy. A model of this type has been calculated by Tubbs (preprint) who achieves an excellent fit to the warped dust lane in Centaurus A 5×10^8 years after the material was captured.

Acknowledgements

I would like to acknowledge many of my colleagues for use of their material, especially K. Hummel, T. van der Hulst, A. Bosma and R. Sancisi. J. Ekers provided the invaluable assistance necessary to produce this manuscript on the required time scale.

REFERENCES

Allen, R.J., Baldwin, J.E., Sancisi, R. (1978). Astron. Astrophys. 62, 397.

Allen, R.J., Ekers, R.D., Burke, B.F., Miley, G.K. (1973). Nature, 241, 260.

Baldwin, J.E. (1976). In "The Structure and Content of the Galaxy and Galactic Gamma Rays", p. 206.

Balick, B. & Brown, R.L. (1974) Astrophys.J., 194, 265.

Beck, R. Berkhuysen, E.M., Wielebinski, R. (1980). Nature 283, 272.

Beck, R., Biermann, P., Emerson, D.T., Wielebinski, R. (1979). Astron. Astrophys. 77, 25.

Beck, R., Klein, U. (1980). In IAU Symposium 94.

van Breugel, W.J.M., Schilizzi, R.T., Hummel, E. Kapahi, V.K. (1980). Astron. Astrophys., in press.

Brown, R.L., Lo, K.Y. & Johnston, K.J. (1978). Astron. J., 83, 1594.

de Bruyn, A.G., Crane, P.C., Price, R.M., Carlson, J.B. (1976). Astron. Astrophys. 46, 243.

Condon, J.J., Dressel, L.L. (1978). Astrophys. J. 221, 456.

Crane, P.C. (1977). Ph.D. Thesis, Massachusetts Institute of Technology.

Crane, P.C. (1979) Astron. J., 84, 281.

Disney, M.J., Wall, J.V. (1977). Mon. Not. Roy. Astr. Soc. 179, 235.

Dressel, L.L., Condon, J.J. (1978). Astrophys. J. Suppl. 36, 53.

Ekers, R.D. (1975). In "Structures and Evolution of Galaxies", ed. G. Setti (Reidel, Dordrecht) pp. 217-245.

Ekers, R.D. (1978a). Physica Scripta, 17, 171.

Ekers, R.D. (1978b). I.A.U. Symp. 77 (Reidel, Dordrecht), p. 49, eds. Berkhuysen and Wielebinski.

Ekers, R.D. (1979). In the IAU joint discussion: "Nuclei of Normal Galaxies", Highlights of Radio Astronomy, in press.

Ekers, R.D., Goss, W.M., Schwarz, U.J., Downes, D., Rogstad, D.H. (1975). Astron. Astrophys. 43, 159.

Ekers, R.D., Sancisi, R. (1977). Astron. Astrophys. 54, 973.

Ekers, R.D. & Kotanyi, C.G. (1978). Astron. Astrophys., 67, 47.

Feretti, L., Giovannini, G. (1980). Astron. Astrophys., in press.

Gioia, I.M., Gregorini, L. (1980). Astron. Astrophys. Suppl. 41,329.

Geldzahler, B.J., Kellerman, K.I. & Shaffer, D.B. (1979). Astron. J. 84, 186-188.

van der Hulst, J.M. (1979). In the IAU joint discussion: "Nuclei of Normal Galaxies", Highlights of Radio Astronomy, in press.

Hummel, E. (1980a.). Ph.D. Thesis, University of Groningen.

Hummel, E. (1980b,c). Astron. Astrophys., in press.

Hummel, E. (1980d). Astron. Astrophys. Suppl. 41, 151.

Ilovaisky, S.A., Lequeux, J. (1972). Astron. Astrophys. 20, 347.

Jenkins, C.R., Scheuer, P.A.G. (1980). Mon. Not. Roy. Astr. Soc., in press.

Kellerman, K.I., Shaffer, D.B., Pauliny-Toth, I.I.K., Preuss, E. & Witzel, A. (1976). Astrophys. J., 210, L121.

Klein, U., Emerson, D. (1980). Preprint.

Kotanyi, C.G. (1980). Astron. Astrophys. Suppl., 41, 421.

Kotanyi, C.G., Ekers, R.D. (1979). Astron. Astrophys. 72, L1.

van der Kruit, P.C. (1977). Astron. Astrophys. 59, 359.

van der Kruit, P.C. (1978). In IAU Symposium 77, ed Berkhuysen & Wielebinski (Reidel, Dordrecht).

van der Kruit, P.C., Allen, R.J. (1976). Ann. Rev. Astron.
    Astrophys. 14, 417.
van der Kruit, P.C., Allen, R.J., Rots, A.H. (1977). Astron.
    Astrophys. 55, 421.
Matheson, D.S., van der Kruit, P.C., Brouw, W.N. (1972).
    Astron. Astrophys. 17, 468.
Miley, G.K. (1980). Ann. Rev. Astron. Astrophys., in press.
Ryle, M., Caswell, J.L., Hine, G. & Shakeshaft, J. (1978).
    Nature, 276, 571.
Sancisi, R., Allen, R.J. (1979). Astron. Astrophys. 74, 73.
Sancisi, R., van der Kruit, P.C. (1980). In IAU Symposium 94.
Schmidt, M. (1968). Astrophys. J. 151, 393.
Segalovitz, A., Shane, W.W., Bruyn, A.G. de, (1976). Nature, 264
    222.
Simkin, S.M. (1979). Astrophys. J. 234, 107.
Sramek, R.A.,(1975) Astrophys J. 198, L13.
Stocke, J.T. (1978). Astron. J. 83, 348.
Stocke, J.T., Tifft, W.A., Kaftan-Kassim, M.A. (1978). Astron. J.
    83, 322.

X-RAYS FROM NORMAL GALAXIES AND
CLUSTERS OF GALAXIES

A.C. Fabian

Institute of Astronomy, Cambridge

INTRODUCTION

Our knowledge of the X-ray emission from normal galaxies has rapidly increased since the start of operation of the Einstein X-ray Observatory Satellite (Giacconi et al. 1979). Prior to its launch in 1978, the only normal galaxies detected in X-rays were the Magellanic Clouds, M31 and, of course, our own Galaxy. Now such galaxies are detectable out to distances beyond the Virgo cluster, and detailed maps showing the location of numerous point sources have been made of the nearer galaxies. It appears that most galaxies have an X-ray luminosity of between about 10^{-3} to 10^{-4} of their visual luminosity.

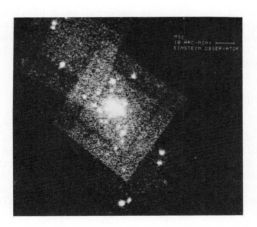

Fig. 1. Three Einstein X-ray Observatory IPC fields of M31. Each field is about 1° in size. The brightest region corresponds to the centre of the galaxy. (van Speybroeck et al. 1979, 1980)

181

The close association of galaxies in clusters and groups of galaxies leads to an enhanced level of X-ray emission (Gursky et al. 1971, Schwartz, Schwarz & Tucker 1980). This is mostly due to thermal bremsstrahlung from a hot intracluster gas. An iron emission feature observed in the X-ray spectrum of many clusters (Mitchell et al. 1976, Serlemitsos et al. 1977) indicates that much of this gas is enriched with heavy elements and suggests that it originated from within the member galaxies themselves. This offers the possibility of studying what may be the total gaseous matter out of which galaxies are made. That which is optically visible often does not directly indicate the gas flows that appear to be part of the life-cycle of a normal galaxy. The X-ray temperature and surface brightness distribution of extended hot gas together probe the gravitational potential of the underlying cluster, or even galaxy, and thus lead to a determination of the mass distribution. Such studies may give valuable insight into the dynamics of galactic systems.

In this review I intend first to summarise the various classes of pointlike and extended X-ray sources in our own Galaxy in order to provide a background to the new results. Some well studied normal galaxies (the Magellanic Clouds, M31, M33, M101 and others) are then covered in some detail, together with the luminosity distribution of their point sources. The first preliminary results of surveys of other galaxies are then reported. Finally the X-ray emission from groups and clusters of galaxies is discussed with an emphasis on the role of the galaxies.

It should be noted at once that few of the major new Einstein Observatory results reported here have yet been published (and some are of a very preliminary nature). These results are the work of others and I rely heavily on reports and discussion generously provided by L. van Speybroeck, K. Long (1980), E. Boldt, A. Rallis and their colleagues.

THE X-RAY STRUCTURE OF OUR GALAXY

Our Galaxy radiates approximately 2×10^{39} erg s^{-1} in the energy range 1-20 keV. This emission is dominated by a few luminous compact X-ray binaries such as GX 5-1, Circinus X-1 and Cygnus X-3. Studies of the X-ray time variability (including in some cases eclipses and pulsations) and the optical counterparts of these sources indicate that accretion of gas onto a compact object (a neutron star or a black hole) is the underlying mechanism. Many of the luminous sources are associated with early-type stars, and as such are probably short-lived ($\sim 10^5$ yr) and may lie along spiral arms. These sources are likely to be found to eclipse. The region within $\sim 30°$ of the Galactic Centre (often termed the 'Bulge' in the X-ray literature), and some globular clusters, also contain luminous compact X-ray sources, which have yet to be seen to eclipse (see review by Lewin & Clark 1980). These are conjectured to be low-mass close binary systems in which the mass transfer may be relatively stable and long-lived ($\gtrsim 10^8$ yr). The most luminous sources in the Galaxy are of this type, and tend to have softer (although sometimes more absorbed) spectra and less variability than the Pop I sources. X-ray bursts are often observed from some members of this predominantly Pop II distribution, although the phenomenon may be more widespread. As yet no X-ray burst has been positively identified with another galaxy, although the 1979 March 7 γ-ray burst may have occurred in N49 in the LMC (Evans et al. 1980). Transient X-ray sources lasting days or even weeks at $> 10^{37}$ erg s^{-1} are relatively common.

Several authors have attempted to define the luminosity distribution of our Galaxy (Seward et al. 1972, Dilworth, Maraschi & Reina 1973, Margon & Ostriker 1973, Markert et al. 1977). Photoelectric absorption in the plane of the Galactic disc may select against particularly soft sources (e.g. in the Einstein Observatory energy band of 0.5-3 keV), and distant sources below $\sim 10^{37}$ erg s^{-1}. An upper limit of a few times 10^{38} erg s^{-1} coincides approximately with the Eddington limit for a solar mass object, and may in future provide us with a distance estimator for other galaxies. This hope is clouded a little, however, by the occurrence of two extremely luminous sources in the Magellanic Clouds; A 0538-68 (Skinner et al. 1980) and SMC X-1 (Clark et al. 1978).

Three other classes of compact galactic X-ray sources are the Galactic Nucleus, cataclysmic variables and pulsars. The Galactic Nucleus was detected by the Einstein Observatory at a level of $\sim 5 \times 10^{35}$ erg s^{-1} (Watson et al. 1980). This region also contains a number of sources. Dwarf novae appear to be relatively weak X-ray sources (Ricketts, King & Raine 1979, Cordova et al. 1980). Only the Crab pulsar has yet been detected as a true X-ray counterpart of a radio pulsar, deriving its energy from rotation and not accretion.

The Einstein Observatory has discovered X-ray emission from stars of all spectral type (Vaiana et al. 1980). The O stars appear to be most luminous at up to 10^{34} erg s^{-1}. The emission mechanism is as yet uncertain. X-ray lines indicate thermal radiation from some stars. Magnetic activity may be directly, or indirectly involved. The high L_x/L_{vis} ratio of up to ~ 0.1 for M dwarf stars allows some members of this class to be detected out to $m_v \sim 17$. It is conceivable that M dwarfs contribute a significant fraction to the diffuse emission from galaxies. Truran (1980) has pointed out to me that a galactic halo composed of low mass stars may potentially be more observable in X-ray detectors than other wavebands. Unfortunately the X-ray data on individual M dwarfs, and the number of M dwarfs in our halo (see e.g. Schmidt 1975 and Lucy 1976) are all sufficiently uncertain that no clear statement can be made at present.

A.C. FABIAN

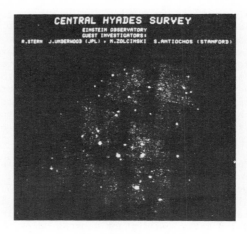

Fig. 2. X-ray emission from many stars in the Hyades cluster.
(Stern et al. 1980)

Several classes of extended X-ray emission may be identified ranging from supernova remnants ($L_x \lesssim 10^{37}$ erg s^{-1}) to the soft X-ray background (L_x (total; $E \lesssim 0.25$ keV) $\simeq 10^{39}$ erg s^{-1} - but very uncertain). O associations (Harnden et al. 1979) and regions such as the Orion Nebula (Ku & Chanan 1979) and η Car (Seward et al. 1979) exhibit complex diffuse emission of total luminosity $L_x \sim 10^{35}$ erg s^{-1}. The soft X-ray background (see Tanaka & Bleeker 1977) is mostly due to hot gas that probably originates from old supernova remnants. Observation of this and identification of any Galactic corona (or wind) is complicated by interstellar absorption and our apparent location in a bubble of hot gas. Ariel V (Warwick, Pye & Fabian 1980) and Uhuru (Schwartz 1980) observations of the 2-10 keV X-ray background indicate a galactic component that ranges from $\sim$ 2-8 percent. This may be due to the inverse Compton scattering of starlight by low energy cosmic ray electrons.

Table 1 summarises the classes of Galactic X-ray source.

THE X-RAY EMISSION FROM NORMAL GALAXIES

The Magellanic Clouds were known to contain about six sources before the launch of the Einstein Observatory. The hundredfold increase in both angular resolution and sensitivity of this new telescope has led to the discovery of many sources in the LMC region (Long, Helfand & Grabelsky 1980). A large concentration of sources is evident around 30 Doradus and at points where star formation seems to be active. 30 Doradus itself appears to be extended by about 80 pc with a luminosity of about 10^{36} erg s^{-1}.

Table 1. X-ray sources in the Galaxy.

Source type	Total Number	$\langle L_x \rangle$ erg s^{-1}	η_x (local) erg s^{-1} pc^{-3}	L_{TOTAL} erg s^{-1}
Early type binaries $L_x > 5 \times 10^{36}$	~ 10	3×10^{37}	–	3×10^{38}
Pop II binaries (?)	~ 20	6×10^{37}	–	1.5×10^{39}
Low luminosity binaries	$\gtrsim 100$	$< 10^{36}$	–	$< 10^{38}$?
Cataclysmic variables	$\sim 10^5$?	$\leq 10^{32}$	$< 10^{26}$	$< 10^{37}$
Galactic Nucleus	–	5×10^{35} E	–	5×10^{35} E
O stars	$\sim 5 \times 10^3$	10^{33} E	–	$\sim 5 \times 10^{36}$ E
Main-sequence stars (B-K)	$\sim 10^{10}$	$\sim 10^{28}$ E	5×10^{26} E	$\sim 10^{38}$ E
M dwarfs halo	$> 10^{11}$?	10^{28} E	$> 10^{25}$ E	10^{39} E
disk	5×10^9			10^{35} E
Supernova remnants	$\sim 10^3$	10^{35}	–	10^{38}
Galactic background E < 0.25 keV	–	–	$\sim 2 \times 10^{28}$ (local)	10^{39}
Galactic background E > 2 keV	–	–	$\sim 10^{26}$	$\sim 10^{38}$

Note: An E in columns 3-5 indicates that the quantity tabulated refers to the Einstein Observatory bandwidth of $\sim$ 0.5 - 3 keV.

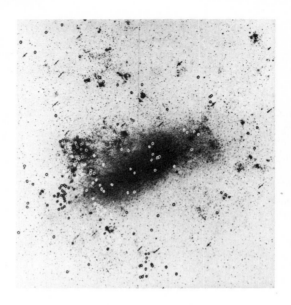

Fig. 3. Positions of X-ray sources found in a preliminary analysis of the Large Magellanic Cloud region (Long et al. 1980). A few of these sources may be foreground stars or background quasars.

Fig. 4. Einstein IPC image of the 30 Doradus region in the Large Magellanic Cloud. LMC X-1 is the strong source at the bottom. The two brightest sources near the centre are N 157B and N 158A. The diffuse emission is from 30 Doradus itself (Long et al. 1980).

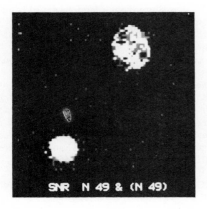

Fig. 5. An HI observation of
N49 and 0527-66 (top right;
Long et al. 1980).

The proximity and low galactic absorption of the Clouds allows
sources down to $L_x \sim 10^{35}$ erg s^{-1} to be found. Many supernova remnants are evident (Long & Helfand 1979) and some are well resolved.
N 132D, for example, is particularly bright in soft X-rays, and
has been found to contain X-ray lines due to iron, silicon and
sulphur (Clark et al. 1980). Most sources have yet to be identified.
The SMC is about an order of magnitude less massive than the LMC,
and a preliminary analysis reveals 26 sources in this region (Seward
& Mitchell 1980).

The Andromeda galaxy (M31) has been observed with both the
Imaging Proportional Counter (IPC with $\sim$ 1.5 arcmin resolution)
and High Resolution Imager (HRI with $\sim$ 2 arcsec resolution) for a
number of 30,000 s exposures. The survey is sensitive to sources

Fig. 6. The 88 X-ray
sources found so far in M31
are here superimposed on the
HRI map of Emerson (1976)
(van Speybroeck et al.1980)

brighter than $\sim 5 \times 10^{36}$ erg s^{-1}. Van Speybroeck and his colleagues (1979, 1980) find 88 sources which distribute along the HI arms, the bulge and nuclear region of the galaxy. The concentration of 19 bright sources apparently within 2' of the nucleus of M31 is significantly higher than that number (3) within a similar region of our Galaxy. The inner sources appear to be about twice as bright as those at the edge of the galaxy. At least 20 sources are identified with globular clusters and appear to be significantly more luminous than those in our Galaxy. Most sources remain unidentified. One source which lies within 3 arcsec of the nucleus of M31 (and may be that nucleus) has been seen to vary over 6-month intervals from 10^{38} to 10^{37} erg s^{-1}. This source accounts for less than 5 percent of the total X-ray emission (at Einstein wavelengths) of M31.

At a similar distance lies M33, which has also been observed several times (Long 1980, Markert et al. 1980). A central source at $\sim 10^{39}$ erg s^{-1} dominates and is more than ten times brighter than any of the eight other sources seen. At least two of these sources are identified with HII regions, and thus presumably with star formation.

A wide range of other normal galaxies has been observed. M101 is ten times further away than M31 or M33 and thus only the tip of

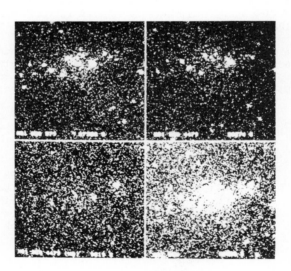

Fig. 7. Three separate HRI observations of the nuclear region of M31. A source positionally coincident with the nucleus is seen to vary. A composite image is shown at bottom right (van Speybroeck et al. 1980).

its luminosity function is observed. Several sources are detected.

There does not at present appear to be any outstanding difference between the Einstein X-ray luminosities of irregular, spiral or elliptical galaxies. The dominant correlation appears at present to be with visual magnitude (Long 1980). Most galaxies emit $\sim 10^{-3}$ to 10^{-4} of their total luminosity as X-rays. The brighter ellipticals studied so far are in the Virgo cluster (Forman et al. 1979, 1980a). They are sufficiently distant that individual point sources are not seen although some are clearly extended. At a luminosity of up to 10^{41} erg s^{-1}, over 10^3 X-ray binaries are required, and it seems reasonable to suppose that a large fraction of their X-ray emission is due to hot gas trapped in their potential wells. Such gas is likely to have been shed by stars and may represent the inner regions of a galactic corona or a wind frustrated by the

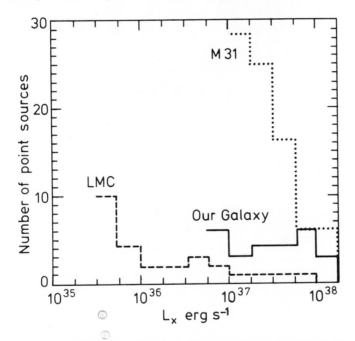

Fig. 8. Approximate luminosity distributions for point X-ray sources in our Galaxy (Markert et al. 1977), M31 (van Speybroek et al. 1980) and the LMC (Long et al. 1980). L_x refers to the 2-11 keV band for our Galaxy, and the 0.5-4.5 keV band for the LMC and M31. This, together with the greater mass at M31, may explain the large difference between the numbers in our Galaxy and those in M31. Note that most of the luminous sources in our Galaxy and M31 are of this 'bulge' type, whereas those in both Magellanic Clouds ($L_x > 3 \times 10^{36}$ erg s^{-1}) are probably Pop I X-ray binaries.

189

surrounding intracluster medium. The level at which diffuse emission contributes to galaxies of different L_{vis}, and Hubble type, is uncertain at present. Bregman (1980) estimates that galactic coronae, or winds, will be difficult to detect from galaxies such as our own. As I shall discuss later, it is possible to accumulate significant masses of gas in a galaxy at X-ray temperatures, which would other-. wise be undetectable.

Fig. 9.   X-ray sources in the region of M101 (Long 1980).

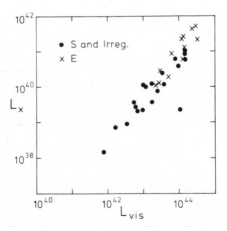

Fig. 10.   The Einstein X-ray luminosity of a sample of irregular, spiral and elliptical galaxies plotted against their respective visual luminosity. Several weak, or undetected, galaxies are not included (Long 1980).

190

Normal galaxies are not significantly brighter than our own
Galaxy at energies above the Einstein range. Worrall, Marshall &
Boldt (1979) failed to detect 76 galaxies (which happened on average
to be less optically luminous than our Galaxy), within about 20 Mpc,
and concluded that they must radiate less than $\sim 10^{39}$ erg s^{-1} in
the 2-10 keV range (their detector on HEAO-1 is sensitive up to
$\sim$ 60 keV). This means that normal galaxies contribute less than
$\sim$ 1 percent to the X-ray background (see also Rowan-Robinson &
Fabian 1975, and van Paradijs 1978). Boldt (1980) finds from scans
of the local supercluster and the luminosity function for active
galaxies that any possible enhancement in that direction is account-
able from an expected supercluster increase in unresolved active
galaxies (perhaps $\sim$ 50) and from normal galaxies compatible with the
findings of Worrall et al. (1979).

The Einstein Observatory results are, of course, showing a
complete range of nuclear activity from the 5 x 10^{35} erg s^{-1} of our
Galaxy up to $\sim 10^{46}$ erg s^{-1} for luminous quasars. However, even an
abnormal galaxy such as Cen A (NGC 5128) has what may be 'normal'
X-ray sources present. An HRI image shows, apart from the nucleus
and extensive jet, a line of emission ($L_x \sim 2$ x 10^{39} erg s^{-1}) just
South of the dust lane (Schreier et al. 1980). A considerable
amount of diffuse emission (not clearly associated with the radio
emission) is also seen.

THE X-RAY EMISSION FROM GROUPS AND CLUSTERS OF GALAXIES

Normal galaxies tend to cluster together in a wide spectrum
of associations ranging from loose groups to rich clusters. The

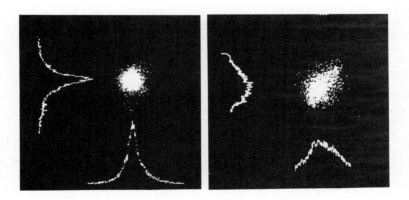

Fig. 11. Two clusters of galaxies observed in the IPC. A85 is to
the left; A1765 to the right. The bright knot to the South in
A1765 is coincident with a radio galaxy (Jones et al. 1979).

X-ray emission from the denser systems significantly exceeds that obtained by summing the separate contributions of the member galaxies. L_x/L_{vis} for the Coma cluster, for example, is $\sim 2 \times 10^{-2}$; which is a factor of 20 higher than that observed for the most luminous individual galaxies.

It is well known that the X-ray emission from rich clusters is dominated by thermal bremsstrahlung (and line-) radiation from a hot gas at temperatures in the range of $\sim 2 \times 10^7$ to 10^8 K. HEAO-A2 and Einstein Observatory X-ray spectra of these clusters confirms that much of this gas is enriched in heavy elements such as iron, silicon, sulphur and oxygen (Mushotzky & Smith 1980, Canizares et al. 1979). It is reasonable to assume that much of this intracluster gas is secondary, and originated from the member galaxies in the past (see e.g. De Young 1978, Hartwick 1980), especially if supernova activity was higher at earlier epochs. The exact means by which the interstellar gas escaped (or was forced) into the intracluster medium is not yet known in detail, although galactic winds (Mathews & Baker 1971), evaporation (Cowie & Songaila 1977) and ram pressure stripping (Gunn & Gott 1972) have all been suggested and may all be relevant. One point that does emerge is that the intracluster gas, which is inferred to have a similar mass to the optically luminous mass in clusters (assuming a missing mass of a factor of ten) must be accounted for in general schemes for the evolution of all galaxies. It is also clear that clusters must evolve in a manner such that much of this gas remains bound to the cluster.

A considerable fraction of the mass shed by stars in isolated ellipticals and other galaxies, is probably lost via a galactic wind (Johnson & Axford 1971, Mathews & Baker 1971). Supernova heating of the interstellar gas, which is mostly lost from giant stars and planetary nebulae, can easily raise its temperature to above $\sim 5 \times 10^6$ K such that mass outflow occurs. The collision of galactic winds with each other and with trapped gas in the increased potential well of an association of galaxies may lead to a situation in which a considerable fraction of this gas resides within a few 100 kpc. The temperature of this gas (if no other heat sources are present) may then be determined by the collision velocities, i.e. the random galaxy velocities of the galaxies, and will be greater in a system with more galaxies than in a sparse one. Radiative cooling may ensue where the gas densities are relatively high and possibly the temperature low. The individual galaxies may each supply ~ 1 M_o yr^{-1} in fresh gas (by evaporation or stripping). This may result in a cycling of gas with perhaps a slower moving central galaxy growing at the expense of the outer ones. Such a situation may be observed in some of the groups and poor clusters recently discovered (Schwartz et al. 1980, Kriss et al. 1980). It may be noted that the cooling time of diffuse gas at 2×10^7 K distributed over 10 R_{10} kpc and detected by the Einstein Observatory is $\sim 5 \times 10^9$ R_{10} $^{\frac{1}{2}}$yr. Thus the smaller and/or brighter

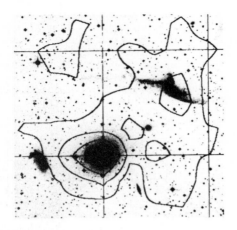

Fig. 12. X-ray emission contours of the Pavo group. The brighter galaxy is NGC 6876. The group is approximately 80 Mpc distant $(H_o = 50$ km s^{-1} Mpc$^{-1})$.

a region is, the more likely it is to cool. Other groups, such as that in Pavo, are weak enough to perhaps be explainable by summation of the emission from their component galaxies; others by some activity of one or more member.

Several very distant rich clusters have been detected by the Einstein Observatory (Henry et al. 1980). There is no indication of any extreme evolution out to a redshift $z = 1$. The current analysis suggests that clusters there are less luminous than those nearby. Several clusters that exhibit the optical effect found by Butcher & Oemler (1978) in which centrally condensed distant clusters contain a large population of galaxies with the colours of spirals, have been observed. The 3C 295 cluster has been resolved by the HRI (Henry et al. 1980) and shows emission predominantly in the region where the redder galaxies are found. This might be expected if they have lost their gas to form the intracluster medium. The early evolution of clusters and galaxies from an X-ray viewpoint has been recently discussed by, among others, Perrenod (1980) and Bookbinder et al. (1980).

Ram pressure stripping may be an important mechanism by which gas is lost from some cluster galaxies, (see numerical studies in Gisler 1976, Lea & De Young 1976, Toyama & Ikeuchi 1980). The effectiveness of this process at early epochs may be reduced by a high gas production rate (Gisler 1979) or by radiative cooling of the dense interstellar gas (Norman & Silk 1979, Sarazin 1979). It may be enhanced by the disruption of massive galactic haloes (the 'missing mass'?).

M86 (NGC 4406) in the Virgo cluster may be one example in which we see ram-pressure stripping in action (Forman et al. 1979, 1980a, Fabian et al. 1980). Here the Einstein observations reveal a distorted plume of X-ray emission. The galaxy is not a known radio source and it appears reasonable to assume that the emission is bremsstrahlung from 1-6 x 10^9 $M_\odot$ of gas at $\sim$ 2.5 x 10^7 K. One obvious difference between M86 and its otherwise similar neighbour M84, which is a much weaker X-ray source with no plume, lies in its radial velocity. M86 is crossing the Virgo cluster at least at 1500 km s^{-1}, which is $\sim$ $\sqrt{2}$ faster than the central velocity dispersion. It may thus be on an orbit bound to the cluster, but not to the core, through which it passes about every 5 x 10^9 yr. In that time it may accumulate $\sim$ 5 x 10^9 $M_\odot$ of its own gas which is only now being stripped away by the increased density of intracluster gas in the Virgo core. The means of accumulating such a mass of gas may not be obvious, for if it is at a temperature less than $\sim$ 2 x 10^7 10^7 K it cools and if hotter and unimpeded it flows out as a wind. Fabian, Schwarz & Forman (1980) suggest that the tenuous medium away from the core of the Virgo cluster is sufficient to confine such a wind and allow much of the gas to follow the orbit of M86, until it reaches the cluster core.

In a more evolved system, the dense gas in the central parts of a cluster may undergo radiative cooling and be driven into smaller radii by the pressure of the surrounding gas (Cowie & Binney 1977, Fabian & Nulsen 1977, Mathews & Bregman 1978). Cooling instabilities in such a flow may lead to optical filamentation (see e.g. Cowie, Fabian & Nulsen 1980). X-ray (Fabian et al. 1980) and optical (Kent & Sargent 1978) observations of the region around NGC 1275 in the Perseus cluster suggest that this process may indeed be taking place there. The accretion rate is $\sim$ 200 $M_\odot$ yr^{-1}.

The intracluster gas must serve as a marker for the cluster (and galactic) potentials. This may be exploited to determine the mass of the cluster, or galaxy. Fabricant, Lecar & Gorenstein (1980) have attempted to measure the mass distribution around M87 by combining X-ray surface brightness observation with several temperature profiles. If the gas is isothermal, the roughly power-law X-ray surface brightness leads to a gravitational mass increasing linearly with radius. The detailed mass profile depends upon the detailed temperature profile, which is not yet well-determined observationally. The present analysis indicates a mass for M87 within a factor of $\sim$ 2 of 10^{13} $M_\odot$ within about 150 kpc.

X-ray surface brightness and temperature distributions are not yet known over any cluster in sufficient detail to allow the mass to be determined. Preliminary work on the Perseus cluster, for which much of the gas has a temperature of 8 x 10^7 K (Mushotzky & Smith 1980), does, however, already suggest a serious discrepancy with the optical data (Gorenstein et al. 1978, Nulsen & Fabian 1980). The high line of sight velocity dispersion of 1420 km s^{-1} would

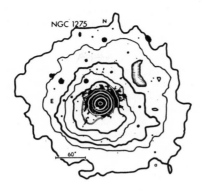

Fig. 13. HRI map of the region surrounding NGC 1275 in the Perseus cluster of galaxies superimposed on an Hα photo by Lynds (1970). A point source is coincident with the nucleus of NGC 1275 (see Fabian et al. 1980).

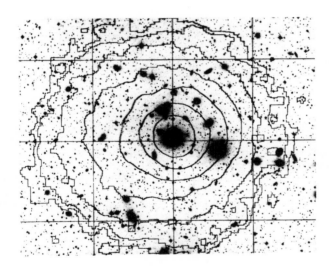

Fig. 14. IPC surface brightness contour map of the Perseus cluster (see Branduardi-Raymont et al. 1980). Notice the asymmetrical elongation of the outer contours. No correction has been made for vignetting etc. at the edge of the field of view.

195

suggest a much tighter X-ray distribution than is observed. This problem may be resolved by considering mainly radial motions and/or the existence of one or more subclusters superimposed on the main cluster.

The existence of subclusters is expected in most theories of cluster formation (see e.g. White 1976). Einstein observations of a sample of $\sim$ 50 clusters do reveal about four of them to be bimodal (Forman et al. 1980b), and perhaps about to merge. The timescale for the separate components in these clusters to intersect on radial orbits is only $\sim$ 5 x 10^8 yr. A statistical study of X-ray subclusters may indicate the timescale on which two subclusters coalesce and provide insight into the formation of rich clusters.

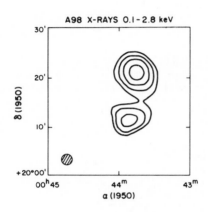

Fig. 15. The two peaks of X-ray emission from A98 (Forman et al. 1980b).

SUMMARY

It is clear that the X-ray astronomy of normal galaxies and clusters is a rapidly growing field. Much of the information on point sources in other galaxies is going to be directly of use for the study of the evolution of, and mechanisms in, binary X-ray sources. For example, the true nature of the 'bulge' sources may be found by observing galaxies other than our own. There are already more sources known in M31 that we knew in our own Galaxy two years ago. The differences in the luminosity function of point sources between galaxies is going to require much explanation. The varied globular cluster populations are intriguing.

Studies of the diffuse X-ray emission associated with hot gas in galaxies are in their infancy, but offer interesting prospects. This gas may be the most enriched, since it originated in the most recent supernova, and its flow pattern may decide (and may have decided) some important aspects of the evolution of a galaxy. In elliptical galaxies, X-ray observations may be the only way to detect much of this gas. Most of the gas shed by galaxies in clusters is still there to be studied for its own sake, as well as being a means of probing the underlying potential well.

ACKNOWLEDGEMENTS

I am indebted to E. Boldt, W. Forman, R. Giacconi, P. Henry, K. Long, T. Markert, P. Nulsen, K. Pounds, A. Rallis, E. Schreier, G. Stewart, J. Underwood and L. van Speybroeck for correspondence and discussion. The Radcliffe Trust is thanked for their support.

REFERENCES

Boldt, E.A. (1980). Private communication.
Bookbinder, J., Cowie, L.L., Krolik, J.H., Ostriker, J.P. & Rees, M.J. (1980). Astrophys.J., 237, 647.
Branduardi-Raymont, G., Fabricant, D., Gorenstein, P., Grindlay, J., Soltan, A. & Zamorani, G. (1980). Astrophys.J. submitted.
Bregman, J.N. (1980). Astrophys.J., 237, 681.
Butcher, H. & Oemler, A. (1978). Astrophys.J., 219, 18.
Canizares, C.R., Clark, G.W., Markert, T.H., Berg, C., Smedira, M., Bardas, D., Schnopper, H. & Kalata, K. (1979). Astrophys.J., 234, L33.
Clark, D. et al. (1980) in preparation.
Clark, G., Doxsey, R., Li, F., Jernigan, J.G. & van Paradijs, J. (1978). Astrophys.J., 221, L37.
Cordova, F. et al. (1980). Astrophys.J. submitted.
Cowie, L.L. & Binney, J. (1977). Astrophys.J., 215, 723.
Cowie, L.L., Fabian, A.C. & Nulsen, P.E.J. (1980). Mon.Not.R.Astr. Soc., 191, 399.
Cowie, L.L. & Songaila, A. (1977). Nature, 266, 501.
De Young, D.S. (1978). Astrophys.J., 223, 47.
Dilworth, C., Maraschi, L. & Reina, C. (1973). Astron.Astrophys., 28, 71.
Emerson, D.T. (1976). Mon.Not.R.Astr.Soc., 176, 321.
Evans, W.D. et al. (1980). Astrophys.J. in press.
Fabian, A.C., Hu, E.M., Cowie, L.L. & Grindlay, J. (1980). Astrophys.J. submitted.
Fabian, A.C. & Nulsen, P.E.J. (1977). Mon.Not.R.Astr.Soc., 180, 479.
Fabian, A.C., Schwarz, J. & Forman, W. (1980). Mon.Not.R.Astr.Soc., 192, 135.
Fabricant, D., Lecar, M. & Gorenstein, P. (1980). Astrophys.J. submitted.
Forman, W., Schwarz, J., Jones, C., Liller, W. & Fabian, A.C. (1979). Astrophys.J., 234, L27.
Forman, W. et al. (1980a). Astrophys.J. submitted.
Forman, W., Bechtold, J., Blair, W., Giacconi, R., van Speybroeck, L. & Jones, C. (1980b). Astrophys.J. submitted.
Giacconi, R. et al. (1979). Astrophys.J., 230, 540.

Gisler, G. (1976). Astron.Astrophys., 51, 137.

Gisler, G. (1979). Astrophys.J., 228, 385.

Gorenstein, P., Fabricant, D., Topka, K., Harnden, F.R. & Tucker, W.H. (1978). Astrophys.J., 224, 718.

Gunn, J.E. & Gott, J.R. (1972). Astrophys.J., 176, 1.

Gursky, H., Kellogg, E., Murray, S., Leong, C., Tananbaum, H. & Giacconi, R. (1971). Astrophys.J., 167, L81.

Harnden, F.R., Branduardi, G., Elvis, M., Gorenstein, P., Grindlay, J., Pye, J.P., Rosner, R., Topka, K. & Vaiana, G.S. (1979). Astrophys.J., 234, L51.

Hartwick, F.D.A. (1980). Astrophys.J., 236, 754.

Henry, P., Branduardi, G., Briel, U., Fabricant, D., Feigelson, E., Murray, S., Soltan, A. & Tananbaum, H. (1979). Astrophys.J., 234, L15.

Henry, P. et al. (1980). Astrophys.J. submitted.

Johnson, H.E. & Axford, W.I. (1971). Astrophys.J., 165, 381.

Jones, C., Mandel, E., Schwarz, J., Forman, W., Murray, S.S. & Harnden, F.R. (1979). Astrophys.J., 234, L21.

Kent, S.M. & Sargent, W.L.W. (1978). Astrophys.J., 230, 667.

Ku, W.H.M. & Chanan, G.A. (1979). Astrophys.J., 234, L59.

Kriss, G.A., Canizares, C.R., McClintock, J.E. & Feigelson, E.D. (1980). Astrophys.J., 235, L61.

Lea, S.M. & De Young, D.S. (1976). Astrophys.J., 210, 647.

Lewin, W.H.G. & Clark, G. (1980). Ann. N.Y. Acad.Sci., 336, 451.

Long, K.S. (1980). Report of talk given at 156th meeting of the American Astronomical Society.

Long, K.S. & Helfand, D.J. (1979). Astrophys.J., 234, L77.

Long, K.S., Helfand, D.J. & Grabelsky, D. (1980) Astrophys.J. submitted.

Lucy, L. (1976). Astrophys.J., 203, 75.

Lynds, C.R. (1970). Astrophys.J., 159, L151.

Margon, B. & Ostriker, J. (1973). Astrophys.J., 186, 91.

Markert, T.H., Canizares, C.R., Clark, G.W., Hearn, D.R., Li, F.K., Sprott, G.F. & Winkler, P.F. (1977). Astrophys.J., 218, 801.

Markert, T.H., Kriss, G.A., Canizares, C.R., McClintock, J.E., Winkler, P.F. & Rallis, A. (1980). Preprint.

Mathews, W.G. & Baker, J.C. (1971). Astrophys.J., 170, 241.

Mathews, W.G. & Bregman, J.N. (1978). Astrophys.J., 224, 308.

Mitchell, R.J., Culhane, J.L., Davison, P.J.N. & Ives, J.C. (1976). Mon.Not.R.Astr.Soc., 176, 298.

Mushotzky, R.F. & Smith, B.W. (1980). In Highlights of Astronomy, I.A.U. publications, Reidel.

Norman, C. & Silk, J. (1979). Astrophys.J., 233, L1.

Nulsen, P.E.J. & Fabian, A.C. (1980). Mon.Not.R.Astr.Soc., 191, 887.

Perrenod, S. (1980). Astrophys.J., 236, 373.

Ricketts, M.J., King, A.R. & Raine, D.J. (1979). Mon.Not.R.Astr. Soc., 186, 233.

Rowan-Robinson, M. & Fabian, A.C. (1975). Mon.Not.R.Astr.Soc., 170, 199.

Sarazin, C. (1979). Astrophys.Letters, 20, 93.

Schmidt, M. (1975). Astrophys.J., 202, 22.

Schreier, E. et al. (1980). Astrophys.J. submitted.

Schwartz, D.A. (1980). Physica Scripta, 21, 644.

Schwartz, D.A., Schwarz, J. & Tucker, W. (1980). Astrophys.J., 238, L59.

Serlemitsos, P.J., Smith, B.W., Boldt, E.A., Holt, S.S. & Swank, J.H. (1977). Astrophys.J., 211, L63.

Seward, F., Burginyon, G., Grader, R., Hill, R. & Palmieri, T. (1972). Astrophys.J., 178, 131.

Seward, F.D., Forman, W.R., Giacconi, R., Griffiths, R.E., Harnden, F.R., Jones, C. & Pye, J.P. (1979). Astrophys.J., 234, L55.

Seward, F.D. & Mitchell, M. (1980). Astrophys.J. submitted.

Skinner, G. et al. (1980). Astrophys.J. submitted.

Stern, R., Underwood, J., Zolcinski, M. & Antiochos, S. (1980). Astrophys.J. submitted.

Tanaka, Y. & Bleeker, J.A.M. (1977). Space Sci.Rev., 20, 815.

Toyama, K. & Ikeuchi, S. (1980). Preprint.

Truran, J. (1980). Private communication.

Vaiana, G.S. et al. (1980). Astrophys.J. in press.

van Paradijs, J. (1978). Astrophys.J., 226, 586.

van Speybroeck, L., Epstein, A., Forman, W., Giacconi, R., Jones, C., Liller, W. & Smarr, L. (1979). Astrophys.J., 234, L45.

van Speybroeck, L. et al. (1980). Astrophys.J. submitted.

Warwick, R.S., Pye, J.P. & Fabian, A.C. (1980). Mon.Not.R.Astr. Soc., 190, 243.

Watson, M., Willingale, R. et al. (1980). Preprint.

White, S.D.M. (1976). Mon.Not.R.Astr.Soc., 177, 717.

Worrall, D.M., Marshall, F.E. & Boldt, E.A. (1979). Nature, 281, 127.

THE GALAXIES OF THE LOCAL GROUP

Sidney van den Bergh

Dominion Astrophysical Observatory

1. GROUP MEMBERSHIP

A census of the 29 galaxies that are presently known to be situated within 1.5 Mpc is given in Table 1. According to arguments presented by Yahil, Tammann and Sandage (1977) objects within this distance are dynamically associated with the Local Group. The data in Table 1 supplement those given in van den Bergh (1979a,b). The object LGS 2 (Kowal, Lo and Sargent 1978) has not been included in the table since I could not see it on limiting IIIaJ and IIIaF plates obtained with the Palomar 1.2m Schmidt telescope last year. The reality of this object needs to be confirmed on large reflector plates. Also excluded from the table is the Phoenix dwarf system (Schuster and West 1976); the status of which is still unclear. Finally NGC 300, which has a distance, derived from Cepheids, of 1.6 (+0.4,−0.3) Mpc (Graham 1979) has been excluded from the table on dynamical grounds (cf. Yahil, Tammann and Sandage 1977). IC5152 has tentatively been retained as a probable Local Group member because of its low radial velocity even though Sérsic and Cerruti (1979) have recently advocated a distance of $\sim$ 4.6 Mpc to this object.

2. DISTRIBUTION OF GROUP MEMBERS

The distribution of Local Group members on the sky is shown in Figure 1. The most striking feature of this figure is the strong concentration near M31. A noteworthy feature of the core of the Andromeda subgroup is that all objects within $10°$ (corresponding to $\sim$ 100 kpc) of M31 are either ellipticals or dwarf spheroidals. Einasto (1978) has previously drawn attention to the fact that close companions to giant galaxies are generally of early type. The Magellanic Clouds, which may however be near perigalacticon of a very elongated orbit (Toomre 1972), are an apparent exception to this rule.

The galactic subgroup of the Local Group, which comprises the LMC, SMC and the Carina, Draco, Sculptor and Ursa Minor systems, is not apparent in Figure 1 because we are so close to its centre.

Table 1

Data on probable Local Group members

Name	α	1950 δ	Type	M_V
M31=NGC 224	$00^h 40^m.0$	$+41°00'$	SbI-II	-21.1
Galaxy	17 42.5	-28 59	Sbc	-20.5:
M33=NGC 598	01 31.1	+30 24	ScII-III	-18.9
LMC	05 24	-69 50	IrIII-IV	-18.5
IC 10	00 17.6	+59 02	IrIV	-17.6
SMC	00 51	-73 10	IrIV/IV-V	-16.8
M32=NGC 221	00 40.0	+40 36	E2	-16.4
NGC 205	00 37.6	+41 25	E6p	-16.4
NGC 6822	19 42.1	-14 53	IrIV-V	-15.7
NGC 185	00 36.1	+48 04	dE0	-15.2
NGC 147	00 30.4	+48 14	dE4	-14.9
IC 1613	01 02.3	+01 51	IrV	-14.8
WLM=DDO 221	23 59.4	-15 44	IrIV-V	-14.7
Fornax	02 37.5	-34 44	D Sph	-13.6
Leo A=DDO 69	09 56.5	+30 59	IrV	-13.6
IC 5152	21 59.6	-51 32	IrIV/IV-V	-13.5:
Pegasus=DDO 216	23 26.1	+14 28	IrV	-13.4
Sculptor	00 57.5	-33 58	D Sph	-11.7
And I	00 42.8	+37 46	D Sph	-11:
And II	01 13.6	+33 11	D Sph	-11:
And III	00 32.7	+36 14	D Sph	-11:
Aquarius=DDO 210	20 44.1	-13 02	Ir	-11:
Leo I=DDO 74	10 05.8	+12 33	D Sph	-11.0
Sagittarius	19 27.1	-17 47	Ir	-10:
Leo II=DDO 93	11 10.8	+22 26	D Sph	-9.4
Ursa Minor=DDO 199	15 08.2	+67 18	D Sph	-8.8
Draco=DDO 208	17 19.4	+57 58	D Sph	-8.6:
Carina	06 40.4	-50 55	D Sph	...
Pisces=LGS3	00 01.2	+21 37	Ir	-8.5:

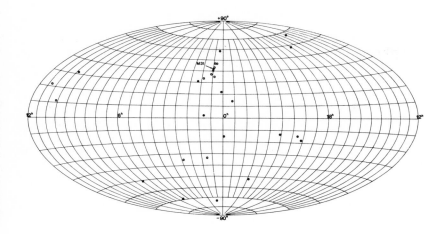

Fig. 1. Distribution of Local Group members on the sky. E and DSph galaxies are shown as open circles.

3. THE DWARF SPHEROIDAL GALAXIES

Dwarf irregular and dwarf spheroidal galaxies are the most numerous constituents of the Local Group. It seems likely that these two types of faint objects are also the most common kinds of galaxies in the Universe. Within the Local Group the dwarf irregulars are mainly situated in a low-density environment whereas the dwarf spheroidals favour the neighborhood of M31 (And I, And II, And III) or the Galaxy (Car, Dra, ScI, UMi). In this respect the Local Group mirrors the Universe at large in which early-type galaxies are concentrated in clusters and late-type galaxies occur mostly in the Field.

A plausible speculation would appear to be that dwarf spheroidals near giant galaxies were originally dwarf irregulars that were stripped of their gas by ram pressure exerted by the gaseous halos of their parent galaxies. Such a scenario might account for the presence of what appear to be intermediate-age objects such as carbon stars (Mould and Aaronson 1979, Renzini and Voli 1980) and anomalous Cepheids (Norris and Zinn 1975, Demarque and Hirshfeld 1975) in dwarf spheroidals. A difficulty with this hypothesis is, however, that it predicts that dwarf spheroidals should be flattened disk-like objects. This is not the case. For the 10 DSph systems in the Local Group the mean axial ratio $<b/a> = 0.75$, which does not differ significantly from the value $<b/a> = 0.77$ for ellipticals. The conclusion that dwarf spheroidal galaxies and ellipticals have similar flattenings is also supported by Karachentseva (1970) who obtained <b/a> = 0.79±0.02 for 89 DSph galaxies which she was able to identify on the prints of the Palomar Sky Survey.

Possibly the existence of intermediate-age objects in some dwarf spheroidals is simply due to the fact that such low-mass objects have long collapse time scales.

4. LOCAL GROUP LUMINOSITY FUNCTION

The integral luminosity functions of rich clusters of galaxies (van den Bergh 1978) all consist of a bright "toe" and a straight line domain at fainter magnitudes. For the straight line region of the luminosity function the number of cluster galaxies brighter than M_F is

$$N = \beta \left[M_F - M_F(0) \right] , \tag{1}$$

in which β is a richness parameter that gives the number of faint galaxies per magnitude interval. For the rich clusters studied by Dressler (1978), $M_F(0)$ has a dispersion of only ~ 0.2 mag (van den Bergh 1978) about the value

$$M_F(0) = -21.15 \pm 0.06 + 5 \log h \tag{2}$$

in which $h = H/100$ km s^{-1} Mpc^{-1}. Transforming the measured (or estimated) V and B-V values of Local Group members to the F system using the relation (Oemler 1974)

$$F = V + 0.35(B-V) - 1.1 \tag{3}$$

yields the integral luminosity function shown in Figure 2. From the data in this figure $M_F(0) = -20.6$ with an estimated uncertainty of ±0.6 mag. Substitution of these values into equation (2) yields $\log h = +0.11 \pm 0.12$; i.e. $H \sim 100$ km s^{-1} Mpc^{-1} is consistent with the data if rich and poor clusters have similar luminosity functions. It would be of interest to obtain $M_F(0)$ values for other nearby clusters such as the M81 group.

5. GLOBULAR CLUSTER FREQUENCY

According to Harris and Racine (1979) the elliptical and dwarf spheroidal galaxies in the Local Group contain a total of 24 globular clusters. The total integrated magnitude (van den Bergh 1979b) of NGC 147, 185, 205, 221 and the Fornax system is $M_V = -17.6$ so that a galaxy of $M_V = -15.0$ made up of Local Group-like ellipticals would contain 2.2 globular clusters. Hereinafter this will be referred to as the specific globular cluster frequency S.

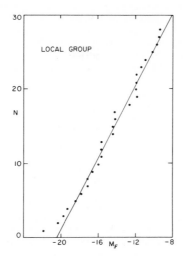

Fig. 2. Luminosity function of Local Group galaxies.

        Assuming the luminosity function of distant globulars to be
similar to that of the Galaxy and the halo of M31 (Racine and Shara
1979) the specific number of globular clusters can be derived for
a number of relatively nearby ellipticals. Provisional results,
partly based on new plate material collected with the CFH 3.6m
telescope in collaboration with W.E. Harris, are given in Table 2.
Fortunately these results are seen to be rather insensitive to the
adopted value of the Hubble parameter h.

Table 2. Specific globular cluster frequencies among elliptical
                                galaxies

Galaxy	h = 1.00	h = 0.75	h = 0.50
<Local Group ellipticals>	2.2	2.2	2.2
NGC 3607	2.7	3.5	...
NGC 4278	13.4	12.0	12.2
NGC 4464	8.3	7.8	12.9
NGC 4472	10.8	10.0	15.6
NGC 5128	0.2±0.8	0.1±0.3	0.04±0.2

The data in the table show that the specific frequency of globular clusters differs significantly from galaxy to galaxy; a fact that was already known from Hanes' (1977) comparative study of the M49 and M87 globular cluster systems. More data will be required to obtain insight into the factors that give rise to the wide variations in specific globular cluster frequency that are observed between different elliptical galaxies. Even within the Local Group striking differences are observed. M32 ($M_v = -16.4$) has S = 0.0 compared to S = 21.8 for the Fornax system ($M_v = -13.6$). Possibly some (but not all cf. van den Bergh 1976) of the low specific globular cluster frequency in M32 can be accounted for by the effects of dynamical friction, which might drag some clusters into the nucleus of this galaxy.

6. DYNAMICAL FRICTION AND GIANT MOLECULAR CLOUDS

In a recent paper Keenan (1979) has studied the effects of dynamical friction on globular clusters with orbits in (or close to) the galactic disk. He concludes that objects with masses of 10^5 $M_\odot$ to 10^6 $M_\odot$ are removed from the disk by dynamical friction on a time-scale that is short compared to the Hubble time. Using numerical techniques Keenan shows that dynamical friction first reduces the peculiar motions of massive objects with respect to the local circular velocity. Subsequently these objects spiral in towards the nuclear bulge of the Galaxy. The arguments given by Keenan for globular clusters apply with equal force to Giant Molecular Clouds (GMC's). According to Solomon and Sanders (1980) the Galaxy contains $\sim$ 4000 GMC's with masses > 1 x 10^5 $M_\odot$. At $\bar{\omega}$ = 5kpc these GMC's account for $\sim$ 90% of the mass of the interstellar medium. With the galactic parameters adopted by Keenan an object with mass M located in the galactic disk will be dragged inwards with a velocity of $\sim$ 1 x 10^{-6} $M/M_\odot$ km s^{-1}. This represents a total inward mass flow in GMC's of $\sim$ 1 $M_\odot$ yr^{-1}. If this phenomenon occurs as predicted it would lead to the destruction of the Galactic disk within a Hubble time unless either (a) the inflow of GMC's is counteracted by an outflow of low-density hydrogen gas or (b) the galactic disk is continuously replenished by infall of $\sim$ 1 $M_\odot$ yr^{-1} of intergalactic gas or (c) tidal friction is less efficient than assumed.

It seems unlikely that the abundance gradient in the galactic disk (Janes 1979) could be maintained in the face of a strong radial circulation in the galactic disk. Since $\sim$ 90% of the gas mass at $\bar{\omega}$ = 5 kpc is in GMC's (which are flowing inwards at $\sim$ 1 km s^{-1}) the low-density hydrogen gas would have to be flowing outwards at $\sim$ 10 km s^{-1} to maintain equilibrium. This might be difficult to reconcile with existing observations (Radhakrishnan and Sarma 1980) of 21 cm line emission in the direction of the galactic center. GMC's with M = 1 x 10^7 $M_\odot$ are calculated to have inward motions of $\sim$ 10 km s^{-1}. Such systematic inward motions should be observable using CO line observations.

Possibly the infall rate of GMC's (and of globular clusters) has been overestimated by Keenan (1979). This could be the case if the drift between kinematical and dynamical LSR (Mihalas and Routly 1968) is much smaller interior to the Sun than it is at $\tilde{\omega}_o$. This might, for example, be the case if the galactic disk contained a zone of constant density at $\tilde{\omega}$ = 3 kpc or if the galactic disk had a central hole (cf. Kormendy 1977, Einasto et al. 1979). This would result in zero dynamical friction locally and hence in a pileup of GMC's at $\tilde{\omega}$ > 3 kpc.

7. GALACTIC DISTRIBUTION OF THE OLDEST OPEN CLUSTERS

In a recent study van den Bergh and McClure (1980) have shown that the oldest known open star clusters are strongly concentrated towards the galactic anti-center. The fact that this phenomenon exists for both low-latitude and intermediate-latitude clusters (see Table 3) shows that it is not due to selection effects resulting from the distribution of absorbing interstellar material.

Table 3.   Galactic longitude distribution of clusters
with ages > 1 x 10^9 yr

Distance from galactic center	All clusters	Clusters $\vert b \vert$ > 10°
0°- 60°, 300°- 360°	1	0
60 - 120 , 240 - 330	8	5
120 - 180 , 180 - 240	11	4

Table 4.   Galactic longitude distribution of the
oldest clusters with T > 3 x 10^9 yr

Distance from galactic center	No. clusters
0°- 60°, 300°- 360°	0
60 - 120 , 240 - 330	2
120 - 180 , 180 - 240	8

Table 4 shows that the concentration towards the galactic anticenter is most pronounced for the oldest (T > 3 x 10^9 yr) clusters. That so few old open clusters are found in the hemisphere centered on the galactic nucleus might well be due to the fact that such clusters frequently meet, and are disrupted by, GMC's which pump energy into them (Spitzer 1958) leading to their eventual disruption. Since GMC's are mostly located interior to the Sun open clusters in the inner part of the galaxy will suffer more disruptive encounters than clusters in the outer part of the galactic disk.

8. THE "CLOUDINESS" OF GALAXIES

It can be shown (van den Bergh and McClure 1980) that the rate at which energy is pumped into open clusters by the Spitzer mechanism is greater if a galaxy contains a few giant GMC's than it is when the same amount of gas is concentrated in many low-mass clouds. The degree to which clusters are disrupted will therefore depend on the "cloudiness" of the interstellar medium. Figures 3 and 4, which have the same resolution since they were both obtained with the Parkes 64m telescope, show the distribution of hydrogen gas in the SMC and in the LMC respectively. The figures clearly show that the interstellar medium in the Small Cloud is distributed much more smoothly than is that in the LMC. Possibly this difference is related to the fact that the SMC gas has a lower metallicity than does that in the LMC. High metallicity and dust content might make it easier for clouds to cool and collapse. Possibly the radial decrease in the Galactic H_2/HI ratio is also due, at least in part, to the metallicity gradient in the galactic disk.

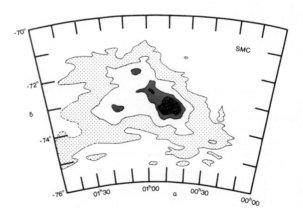

Fig. 3. Distribution of neutral hydrogen (Hindman 1967) in the SMC.

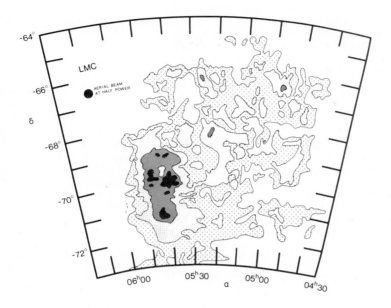

Fig. 4. Distribution of neutral hydrogen (McGee and Milton 1966) in the LMC.

REFERENCES

Demarque, P. & Hirshfeld, A.W. (1975). Astrophys.J., 202, 346.
Dressler, A.M. (1978). Astrophys.J., 223, 765.
Einasto, J. (1978). In The Large Scale Structure of the Universe
     I.A.U. Symposium No. 79, eds. M.S. Longair and J. Einasto,
     p. 51 (Reidel Publ.Co. - Dordrecht).
Einasto, J., Tenjes, P., Barabanov, A.V. & Zasov, A.V. (1979).
     Tartu Preprint A1.
Graham, J.A. (1979). Bull.Am.Astron.Soc., 11, 694.
Hanes, D.A. (1977). Mon.Not.R.Astr.Soc., 180, 309.
Harris, W.E. & Racine, R. (1979). Ann.Rev.Astron.Astrophys., 17,
     241.
Hindman, J.V. (1967). Austr.J.Phys., 20, 147.
Janes, K.A. (1979). Astrophys.J.Suppl., 39, 135.
Karachentseva, V.E. (1970). Problemy Kosmicheskoi Fiziki, 5, 217.
Keenan, D.W. (1979). Astron.Astrophys., 71, 245.
Kormendy, J. (1977). Astrophys.J., 217, 406.
Kowal, C.T., Lo, K-Y. & Sargent, W.L.W. (1978). I.A.U. Circular 3305.

McGee, R.X. & Milton, J.A. (1966). Austr.J.Phys., 19, 343.

Mihalas, D. & Routly, P.M. (1968). Galactic Astronomy, p. 101, (Freeman and Co. - San Francisco).

Mould, J. & Aaronson, M. (1979). Astrophys.J., 232, 421.

Norris, J. & Zinn, R. (1975). Astrophys.J., 202, 335.

Oemler, A. (1974). Astrophys.J., 194, 1.

Oort, J.H. (1970). Astron.Astrophys., 7, 381.

Racine, R. & Shara, M. (1979). Astron.J., 84, 1694.

Radhakrishnan, V. & Sarma , N.V.G. (1980). Astron.Astrophys., 85, 249.

Renzini, A. & Voli, M. (1980). Preprint.

Sandage, A., Freeman, K.C. & Stokes, N.R. (1970). Astrophys.J., 160, 831.

Schuster, H.E. & West, R.M. (1976). Astron.Astrophys., 49, 129.

Sérsic, J.L. & Cerruti, M.A. (1979). Observatory, 99, 150.

Solomon, P.M. & Sanders, D.B. (1980). In Giant Molecular Clouds in the Galaxy, eds. P.M. Solomon and M.G. Edmunds, p. 41, (Pergamon Press - Oxford).

Spitzer, L. (1958). Astrophys.J., 127, 17.

Thuan, T.X. & Martin, G.E. (1979). Astrophys.J.Lett., 232, L11.

Toomre, A. (1972). Q.J.Roy.Astr.Soc., 13, 266.

van den Bergh, S. (1976). Astrophys.J., 203, 764.

van den Bergh, S. (1978). Astrophys.Space Sci., 53, 415.

van den Bergh, S. (1979a). In The Large Scale Characteristics of the Galaxy - I.A.U. Symposium No. 84, p. 577. Ed. W.B. Burton (Reidel Publ. Co., Dordrecht)

van den Bergh, S. (1979b). Mem.Soc.Astron.Italy, 50, 11.

van den Bergh, S. (1979c). Astrophys.J., 230, 95.

van den Bergh, S. & McClure, R.D. (1980). Astron.Astrophys., in press.

Yahil, A., Tammann, G.A. & Sandage, A. (1977). Astrophys.J., 217, 903.

CHEMICAL EVOLUTION OF NORMAL GALAXIES

B. E. J. PAGEL

Royal Greenwich Observatory, Herstmonceux

1. INTRODUCTION AND OVERVIEW

One of the tests of any successful theory of galaxy evolution is that - when combined with what we know (or think we know) about stellar evolution and nucleosynthesis - it should be able to account for what we know (or think we know) about the distribution of the elements and their isotopes in the universe. Conversely, from the observed distribution of elements, we may be able to draw conclusions both about nucleosynthesis and about the collective effects of stellar evolution and the various interactions between stars and the interstellar medium (ISM) in galaxies. These questions form the subject matter of what is usually called galactic chemical evolution, although this name could also be applied to the formation of chemical compounds in the ISM (not discussed here).

The importance of galactic processes in affecting the abundance distribution was already recognised in the classic nucleosynthesis paper by B^2FH (Burbidge et al. 1957) and the subject was developed further in fundamental papers by Eggen, Lynden-Bell and Sandage (1962) and Schmidt (1963). In recent years there has been considerable progress both in the theory of stellar evolution and nucleosynthesis, and in empirical knowledge of element abundances in stars and galaxies, giving rise to numerous papers and reviews. Among the latter I recommend particularly those by Trimble (1975), Audouze and Tinsley (1976) and Tinsley (1980); the present lecture is largely a development of material described by Pagel (1979b).

To fix ideas, Figure 1 shows (very schematically!) a scenario for the evolution of galaxies, stars and elements. The Big Bang leads to significant initial abundances of H, D, ^{3}He, ^{4}He and possibly ^{7}Li in the intergalactic medium (IGM). D is subsequently depleted by stellar activity while ^{4}He is mildly topped up. ^{3}He and ^{7}Li are both created and destroyed in stars. The IGM is a somewhat ghostly entity, whose most tangible manifestation is in X-ray clusters of galaxies where it rather embarrassingly shows an appreciable iron abundance like 1/3 solar; this may be a by-product of conventional enrichment of the ISM and the interaction between galaxies and their environment, and not typical of the IGM in general. Other possible manifestations of the IGM are the absorption-line systems observed shortward of emission Lyman-α in high red-shift QSOs (Sargent et al. 1980) and the effects of inflow into our own or other galaxies that have been invoked for a number of reasons.

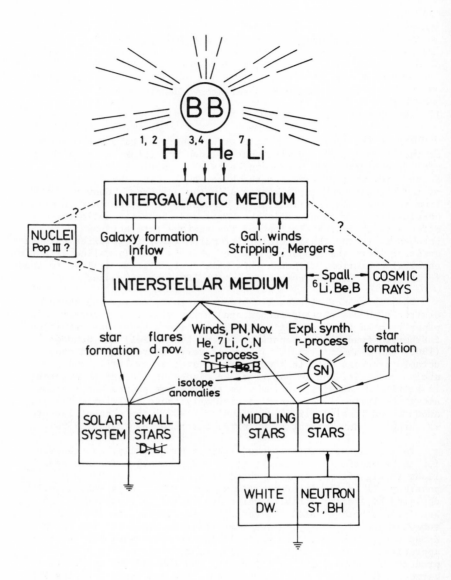

Fig 1. A scenario for cosmic evolution.

The IGM is assumed (for present purposes at least) to condense into galaxies consisting initially of diffuse material which then makes stars either on a short timescale (to make E galaxies and the bulge and halo components of Spirals) or on a long time scale (to make disk components of Spirals and Irregulars, in which substantial quantities of gas are still present). (Alternative scenarios can be envisaged, subject to certain constraints like having to allow abundance gradients to develop in many large galaxies.)

When stars are formed, there are broadly three things that can happen to the ISM as a result of their evolution. Small stars with less than about one solar mass ($M_\odot$) have such long lifetimes that essentially nothing happens (except perhaps minor effects of flares and dwarf novae): they simply serve to lock up a part of the diffuse material and take it out of circulation. Middle-sized stars (say 1 to 10 $M_\odot$) change some of their surface carbon into nitrogen as they evolve up the giant branch and subsequently undergo helium shell flashes in the course of their second ascent, in which helium, lithium, carbon, nitrogen and s-process elements are freshly synthesised to varying degrees and dredged up to the surface as can be seen in carbon and S stars and some planetary nebulae. Other, more mysterious mixing effects can occur even before the second ascent up the red giant branch, as manifested by the barium stars and the variety of anomalous red giants seen in globular clusters. Through stellar winds and planetary nebulae, some or all of the material above the degenerate core is eventually ejected, while the core itself is usually assumed to remain as a white dwarf or neutron star although some stars could simply collapse into black holes and others explode leaving no remnant at all. Above about 10 $M_\odot$ we have the third category of stars, the big ones, which evolve rapidly to a situation where iron is synthesised in the core leading in some cases at least to a supernova outburst in which the products of explosive nucleosynthesis (from carbon up to the iron group and r-process elements) are ejected, sometimes preceded by H and N-rich material, as can be seen in fresh young supernova remnants like Cas A and N 132 D in the Large Magellanic Cloud. The net result of all these processes is to produce a "standard abundance distribution" in our vicinity with typical mass fractions $X \simeq 0.7$ for hydrogen, $Y \simeq 0.28$ for helium and $Z \simeq 0.02$ for heavy elements consisting chiefly of O, C, N, Ne, Mg, Si and Fe.

2. SURVEY OF THE OBSERVATIONAL DATA

The relevant observational data can be classified into two categories: (i) Those relating to stellar masses and luminosities, star formation, galactic dynamics and the morphology, content, mass:light ratio, colours and ages of galaxies – virtually the whole of astronomy; and (ii) data relating more specifically to the

distribution of nuclear species in the universe. It is just the
second category that will be reviewed briefly now, and I shall say
nothing about such important topics as isotopic abundances in the
ISM, isotope anomalies in the solar system, the light elements and
radioactive cosmo-chronology.

2.1 THE STANDARD OR "COSMIC" ABUNDANCE DISTRIBUTION

The prime source of ideas on nucleosynthesis is the abundance
distribution of the Solar System at the time of formation, based on
a combination of analyses of the absorption-line spectrum of the
solar photosphere with direct analysis of elements and isotopes in
meteorites, especially the C1 chondrites. Supplementary data come
from visible, u.v. and X-ray emission lines from the corona, solar
flare particles, the solar wind and occasionally planetary
atmospheres.

Outside the solar system one has a more limited set of data for
young stars, gaseous nebulae (notably the Orion Nebula) and mature
supernova remnants, all of which should represent the present
composition of the ISM, while older stars are taken to represent
the ISM at the time and place of formation except in so far as
their atmospheres have been affected by internal evolution and
mixing. By and large, stellar and interstellar abundances are not
much different from the standard distribution (cf. Table 1), apart
from variations in the overall abundance of heavy elements from
carbon upwards (often briefly referred to as "metals") relative to
hydrogen. There is a rather disturbing discrepancy of about a
factor 2 between the oxygen abundance in Orion and other nearby HII
regions, on one hand, and Sun and hot stars on the other, which may
or may not be real. This discrepancy needs to be borne in mind
when we try to combine data from stars and HII regions to compare
abundances in different parts of the universe.

Table I

"COSMIC" ABUNDANCES

	$12 + \log \frac{He}{H}$	$12 + \log \frac{O}{H}$	$\log \frac{O}{N}$	$\log \frac{O}{Ne}$	$\log \frac{O}{S}$	
PHOTOSPHERE		8.9	0.9		1.7	
CORONA			0.7	1.1	1.6	
PROMINENCE	10.8					
FLARE PTCLS	10.8*		0.9	0.9	1.8	
SOLAR WIND	10.6	8.3		0.5		
B STARS	11.0	8.9	1.0	1.0	(1.5)	
ORION NEB	11.0	8.6	0.9	0.7	1.3	

*From $\overline{(He/O)}$ x (O/H)
FLARE   PHOTOSPHERE

2.2 ABUNDANCES IN STARS

Important features of abundances in stars are (i) the dependence of total heavy-element abundance or "metallicity" on the star's age and galactic orbit, (ii) the distribution function of metallicities (which is a more readily observable consequence of (i)), (iii) abundance anomalies found in stars (and planetary nebulae) in advanced stages of evolution, but sometimes also in not so advanced stages and (iv) differences between abundance ratios of elements (or isotopes) in unevolved stars of different populations, attributable to differential enrichment of the ISM. Historically, effects (i) and (iv) have proved to be the most difficult to establish observationally (cf. Pagel 1979a).

If we take the stars of the Bright Star Catalog, which form a relatively unbiased sample of the solar neighbourhood, and study the distribution function of metallicity (specifically [Fe/H] where [X] denotes the difference between log X in the star and log X in the Sun), we find essentially a Gaussian distribution with a mean near -0.15 and a standard deviation near 0.2. Just 3 stars out of about 10^4 form a separate "halo" population (extreme Pop.II) with much lower metallicity $-2.6 \leqslant$ [Fe/H] $\leqslant -1.3$ and highly eccentric galactic orbits relating them to the globular clusters which also usually have [Fe/H]$\leqslant-1$. Eggen, Lynden-Bell and Sandage (1962) studied the kinematics and metallicities of halo

215

B. E. J. PAGEL

stars and found them to be consistent with a picture of formation
by rapid collapse during which the halo stars were formed before
those of the disk. The idea of enrichment leads us to expect that
the oldest disk stars should be metal-deficient as well and this
indeed seems to be the case, but this has been quite hard to
establish for various reasons: e.g. confusion of the resulting
metallicity-kinematics relation by a radial abundance gradient
increasing towards the centre (Janes 1979), the difficulty in
measuring stellar ages accurately enough and the lack of any very
significant enrichment in the present-day ISM relative to the Sun
(cf. Table 1) and the oldest galactic clusters (which have about
the same age as the Sun). However, by careful studies of the

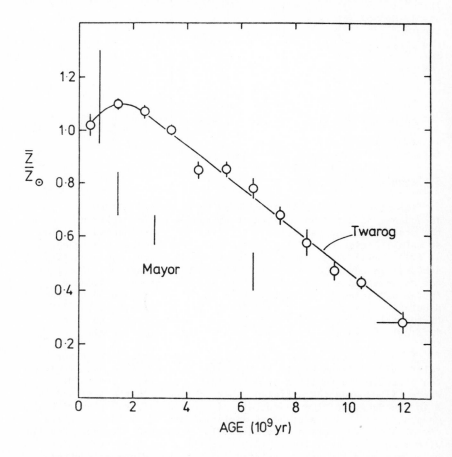

Fig 2.   Relationship between age and mean metallicity for F stars
         in the solar neighbourhood.

Strömgren intermediate-band photometry of field stars of spectral
type F, Mayor (1976) and Twarog (1979) have produced convincing
age-metallicity correlations which are in at least qualitative
agreement (Fig 2) and imply a more or less linear increase of mean
abundance with time, at least up to the birth-time of the Sun.

Superposed on the quasi-uniformity of the heavy-element mixture
are a number of "anomalies" shown by individual stars as a result
of evolutionary processes (Baschek 1979), mainly affecting C, N and
s-process elements, which are probably significant in enriching the
ISM through mass loss by stellar winds and planetary nebulae, which
latter themselves show corresponding anomalies in some cases
(Peimbert 1978; Aitken et al 1979). Red giants of Pop.II show
similar anomalies to a much greater extent. Hotter stars also
show a variety of anomalies, some of which (Bp, Ap and Am stars)
are probably due to diffusion effects at the surface and therefore
merely constitute a red herring as far as galactic evolution is
concerned, but others (Wolf-Rayet stars, close binaries) are
associated with strong mass loss and may be highly relevant.

Apart from these "anomalies" associated with evolution of
individual stars, there are smaller population effects suggesting
differential enrichment in different heavy elements (Pagel 1979a).
In metal-deficient stars O and Ca tend to be deficient by smaller
factors than iron while N and heavy s-process elements tend to be
more deficient. Odd-even effects may be present in some cases; e.g.
in Groombridge 1830 with [Fe/H] $\simeq$ -1.3,^{25}Mg and ^{26}Mg are down
relative to ^{24}Mg (Tomkin and Lambert 1980). The only surprising
thing about these differences is that they are so small (cf.
equation 4.7 below); to a first approximation we can consider the
heavy elements to go up or down, relative to hydrogen, as a unit.

Turning to stellar populations in external galaxies (where it
is difficult or in most cases impossible to analyse individual
stars), an indication of the composition of old populations is
given by the colours and the strengths of major spectral features
in integrated light, notably blue CN and MgI + MgH in the green,
combined with plausible population models (Faber 1977). In
ellipticals and the bulges of early-type spirals the mean
metallicity increases with the size of the system, with so-called
super-metalrich stars ([Fe/H] $\simeq$ 0.5) present in the central
population of the largest ones. This loose relationship becomes
considerably tighter when a term increasing with velocity
dispersion is included (Terlevich et al 1980). Furthermore there
are marked abundance gradients in giant Ellipticals and bulges,
with the strongest features occurring close to (but not
necessarily quite at) the centre. An abundance gradient is also
present (though with large scatter) among the globular clusters of
our Galaxy (Harris and Canterna 1979).

2.3 ABUNDANCES IN HII REGIONS AND SUPERNOVA REMNANTS

HII regions are very useful because they enable He, O and N abundances (and in favourable cases those of Ne, S and Ar) to be determined in the interstellar medium in other galaxies and remote parts of our own Galaxy. Mature supernova remnants give somewhat similar information (Dopita, D'Odorico and Benvenuti 1980). In our own Galaxy the prospects are particularly good because optical and radio observations can be combined and the Galactic HII regions reveal a marked abundance gradient in O/H, N/H etc similar to that already mentioned in connection with stars (Peimbert 1979; Talent and Dufour 1979; Mezger et al 1979; Shaver, McGee and Pottasch 1979). In Irregular galaxies like the Magellanic Clouds, on the other hand, there is little or no gradient and for these particular objects it turns out that oxygen is deficient (especially in the Small Cloud) and nitrogen even more deficient (Pagel et al 1978). In Spiral (Scd) galaxies like M33, M101 and NGC 300, the HII regions again reveal a marked abundance gradient, by about an order of magnitude between the central and outermost HII regions (Searle 1971; Smith 1975; Shields and Searle 1978; Pagel et al 1979), but less of a gradient in two barred spirals, NGC1365 (SBbc) and NGC1313 (SBd:Pagel, Edmunds and Smith 1980). A relationship between size (or mass) and oxygen abundance exists for Irregulars (Lequeux et al 1979), similar to the metallicity relation for Ellipticals. However, some Spirals like M83 and M51 have apparently much larger abundances in their outer HII regions (Dufour et al 1980) than other presumably comparably large ones like M101; this may be related to the morphological type, or to the mean surface brightness, but the reasons are not yet clear.

The situation in galactic nuclei themselves is somewhat mysterious. Many normal galaxies have emission-line nuclei with very strong [NII] lines and in M81 and M51 large overabundances of nitrogen have been inferred (Alloin et al 1979), but the mechanism of excitation is still uncertain. In many of the so-called hot-spot galaxies, which have multiple line-emitting spots in the nuclear region, one of the components is turning out to be a Seyfert-like nucleus with X-rays, broad Balmer lines, high excitation and a non-thermal continuum (Véron et al 1980; Ward et al 1980; Wilson, Ward and Blades 1980; Edmunds and Pagel 1980) and the conventional HII region emission (with ionisation by hot young stars) has to be separated from the Seyfert-type emission before one can say anything about abundances.

CHEMICAL EVOLUTION OF NORMAL GALAXIES

3. INGREDIENTS OF CHEMICAL EVOLUTION MODELS

To put together a model for the chemical evolution of galaxies, we need the following ingredients:-

(i)   Initial conditions.

(ii) Some picture of the end-products of stellar evolution, i.e. which stars eject how much of their mass in the form of various heavy elements after how much time?

(iii) The initial mass function (IMF) which gives the relative birthrates of stars of different masses.

(iv) A model of the star formation rate as a function of time, gas mass, gas density and probably other parameters.

(v)   Assumptions about all the other relevant processes in galactic evolution besides the birth and death of stars.

We discuss each of these in turn.

3.1 INITIAL CONDITIONS

Mostly one assumes that we begin with pure gas left over from the Big Bang, but some models appeal to a "prompt initial enrichment" due either to pregalactic synthesis of some kind (doubtful in view of the existence of some stars and some galaxies with very low metal or oxygen abundances indeed) or to a preceding generation of massive stars, which leads to a significant initial abundance of heavy elements.

3.2 END-PRODUCTS OF STELLAR EVOLUTION

In principle these should be predictable from theory but in practice there are many grey areas such as the effects of mixing and mass loss, the precise mechanism and conditions of stellar explosions and the corresponding mass limits between which various things happen. Fig 3 shows a working prescription or "Identikit" set (after Alloin et al. 1979), which is based on evolutionary calculations by Iben, Arnett and others together with some assumptions about mixing effects which help to account for the abundances of carbon and nitrogen. The lowest curve represents the compact residue left over at the end of stellar evolution, white dwarf or neutron star, to which we have to add all the stars born with less than $1M_\odot$ or so whose lifetimes exceed the age of the Galaxy. All the rest of the material is expelled after more or less nuclear processing. Zones 1-3 are H-rich envelope material in which light elements are destroyed and some of the carbon initially present in the star is changed into nitrogen (a

219

secondary process) but otherwise nothing much happens. Zones 4,
5 and 7 are heliumized or carbonized core regions in which N and
fresh carbon are dredged up in He shell flashes and N is produced
both by secondary and by primary processes using the fresh carbon.
Zones 8 and 9 consist of C and O respectively, and the deeper
zones are mainly Ne-Mg and Si-Fe, all four zones being lumped
together for most purposes as the fraction of primary heavy
elements expelled by supernovae.

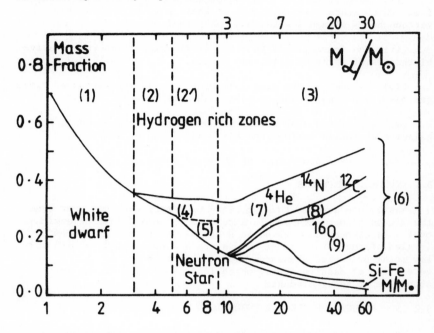

Fig. 3   Model of the end products of evolution of stars of various
masses, after Alloin et al. (1979).

A complication not included in this picture is the effect of
mass loss by massive stars in the core H and He burning phases,
which can have a substantial effect on the yield of heavy elements
produced by a star of given initial mass (Chiosi and Caimmi 1979).
Basically the mass of the products depends on the core mass, M_α,
which is indicated (for the constant-mass case) along the top line
in Fig 3.   If mass loss is appreciable, then the initial stellar
mass for a given M_α has to be increased by a factor of up to 2
or so, the scale of initial masses will be stretched and the yield
of heavy elements from a given stellar population correspondingly
reduced.   Other complications may be caused by the evolution of
close binaries, the effects of which are usually neglected, and
iron synthesis by Type 1 supernovae of intermediate mass.

CHEMICAL EVOLUTION OF NORMAL GALAXIES

3.3 THE INITIAL MASS FUNCTION (IMF)

To go from the results of evolution of single stars to the results from a population of stars, we need to know the relative birthrates of stars with different masses – the IMF (Salpeter 1955). This function is deduced, using the mass–luminosity relation, by comparing the numbers of stars of each mass (in a cylinder perpendicular to the galactic plane) with their calculated evolutionary lifetimes, when these are less than the age of the Galaxy, i.e. $M \geqslant 1.5_\odot$ or so. The segment corresponding to less massive, longer-lived stars is deduced from the corresponding numbers of stars directly (since all those ever born are still here), but to relate this segment to the other segment one needs to make assumptions as to how the rate of star formation (SFR) has varied in the past. Or alternatively, assuming the IMF to be continuous and invariable with time, one can use the condition of a smooth fitting to place constraints on past variations in the SFR. Miller and Scalo (1979) have found in this way that the mean SFR in the past has probably not differed from the present SFR by more than a factor of 2 in either direction; in particular, there are not enough small stars around, compared to big stars, to sustain the view that the rate of star formation has been much greater in the past (cf. also Mayor and Martinet 1977).

From the galactic chemical evolution point of view, we are interested in (a) the total mass that dies within the age of the Galaxy, which we can conveniently identify with the mass fraction ζ that comes in stars exceeding 1 $M_\odot$, and (b) the fraction that comes in stars of a given mass range, say m to m + dm in solar units, with m > 1. This fraction, then is

$$\zeta \, \psi \, (m) \, dm / \int_1^{m_{max}} \psi(m) \, dm \qquad (3.1)$$

for an initial mass function $\psi(m)$. Salpeter adopted for this function a power law

$$\psi \, (m) = Km^{-x} \qquad (3.2)$$

with x = 1.35, but the more modern discussions indicate a curved relationship with x increasing from 0.6 or less below 1.8 $M_\odot$ to 2 above 1.8 $M_\odot$ (Lequeux et al. 1979; Miller and Scalo 1979). ζ is rather uncertain; estimates vary from about 0.2 to about 0.5. Taking x = 2 above 1.8$M_\odot$, we have

$$f(m_1, \, m_2) \simeq 1.2\zeta \, (m_1^{-1} - m_2^{-1}) \qquad (3.3)$$

for the mass fraction of a generation of stars that comes in stars with masses between m_1 and m_2.

3.4 THE RATE OF STAR FORMATION (SFR)

Star formation in disk-like galaxies is a highly sporadic phenomenon, the physics of which is poorly understood and is under active investigation at the present time (e.g. Larson 1977). For our purposes we are interested in the average effects of star formation over large regions and long times, and in this case many people have assumed a simple power-law dependence on volume density, or surface density or simply on the total mass of gas in the system that one is considering:

$$\frac{ds}{dt} = kg^n \qquad\qquad (3.4)$$

where s is the mass in the form of stars and g the mass of gas with n usually considered to be 1 or 2 (Schmidt 1959). The "constant" k may be modified to take into account effects that may affect the overall star formation rate like the frequency of passage through a spiral shock (Jensen, Strom and Strom 1976), metal-enhanced star formation due to the increased cooling efficiency of metal-rich systems (Talbot 1974) and variations in the galactic scale height (Talbot and Arnett 1975; Talbot 1980). An alternative to such laws is simply to assume that ds/dt varies with time in some arbitrarily prescribed way and compute the consequences. We shall see later how to get an idea of the behaviour of the SFR in our own neighbourhood in the past, from the observed age-metallicity and number-metallicity relations, but several of the results of galactic chemical evolution models are not very sensitive to the SFR.

3.5 SUPPLEMENTARY ASSUMPTIONS

The final ingredient of chemical evolution models is a set of assumptions about the conditions under which stellar births and deaths take place. In particular, it makes a considerable difference whether we are allowed to consider an isolated, well-mixed zone or whether inflows and outflows of gas and inhomogeneities such as metal-enhanced star formation are important. Furthermore, it may be necessary to consider systematic variations in the IMF, which are usually neglected. Actually it is almost certain that the IMF varies, with rather more massive (or at least hotter stars) being present in systems of low heavy-element abundance like the Small Magellanic Cloud and the outer regions of Scd galaxies (Shields and Tinsley 1976), but what is less clear is whether such variations have a significant effect on heavy-element synthesis. The answer probably is that they have an influence on certain details like the N/O ratio (Alloin et al. 1979), but their effects on the broad sweep are not yet clearly perceived.

CHEMICAL EVOLUTION OF NORMAL GALAXIES

4. SOME ACTUAL MODELS

To compute chemical evolution, one allows stars to form and evolve according to the appropriate time scales and mixes the ingredients described in Chapter 3 to produce evolutionary abundance curves (e.g. Audouze and Tinsley 1976) describing the abundances of various elements as a function of time, together with the mass of stars, integrated luminosities and colours etc. We shall not go into all these details, preferring to understand as much as we can by simple analytical approximations.

4.1 THE INSTANTANEOUS RECYCLING (IR) APPROXIMATION

A particularly useful approximation, which simplifies everything considerably, is that of instantaneous recycling (Talbot and Arnett 1971; Searle and Sargent 1972), which supposes that everything happens very quickly compared to the timescale of galactic evolution. This is true for explosive synthesis from massive stars, but less true for sythesis of nitrogen and other elements coming from lower-mass stars and for re-ejection of unprocessed material. As a rule of thumb we can regard the assumption as being adequate for products from massive stars as long as most of the gas left over has not been recycled.

In this approximation any assumed combination of IMF and end-products of stellar evolution can be characterised by a few simple parameters, namely (i) the mass fraction of a generation of stars that remains locked up in long-lived stars or compact remnants, which I shall call $\alpha \equiv 1 - \beta \geqslant 1 - \zeta$; and (ii) the mass fraction $q_i \equiv \alpha \, p_i$ that is ejected in processed material in the form of any particular element i. p_i is called the "yield" of the element in question and I shall denote by $p \equiv \Sigma p_i (A \geqslant 12)$ the total yield of heavy elements. Theoretically, p is somewhere in the region of 1 per cent, but with an uncertainty of a factor of 2 or more.

The IR approximation enables us immediately to write down a differential equation for enrichment of the ISM by star formation, including a term allowing for inflow of foreign material with heavy-element abundance Z_f. Let S(t) be the mass of stars ever born up to time t, in a well-mixed system with total mass m (t). Then

$$\frac{d}{dS} (Zg) = q + Z(\beta - q) - Z + Z_f \frac{dm}{dS} \qquad (3.5)$$

where the first term on the R.H.S. comes from ejection of processed material by stars, the second from ejection of unprocessed material (this term would be zero for an element like D which is destroyed in hydrogen-rich zones), the third from heavy elements removed from the medium by star formation and the fourth from inflow (or alternatively outflow with $Z_f = Z$ and $\frac{dm}{dS} < 0$). Replacing S by the actual mass of stars, $s = \alpha S$, this equation reduces to

$$\frac{d}{ds} (Zg) = p(1-Z) -Z +Z_f \frac{dm}{ds}$$

$$\simeq p - Z + Z_f \frac{dm}{ds} \qquad (4.1)$$

or, alternatively,

$$g \frac{dZ}{ds} = p - (Z-Z_f) \frac{dm}{ds}. \qquad (4.2)$$

4.2 THE SIMPLE MODEL AND ITS SUCCESS: IRREGULAR GALAXIES

The Simple or one-zone model (Schmidt 1963; Talbot and Arnett 1971; Searle and Sargent 1972) makes the following assumptions:-

(i)   Evolution takes place in an isolated zone, e.g. the whole of an Irregular galaxy or a cylindrical shell perpendicular to the Milky Way plane, with no inflows or outflows, i.e.

$$m = g(t) + s(t) = const. \qquad (4.3)$$

(ii)   The zone is well-mixed at all times, ie $Z = Z(t)$ only.

(iii) We begin with pure gas with no heavy elements, i.e.
$g(0) = m$ and $Z(0) = 0$.

(iv)   The yield is constant.

With these assumptions we can substitute $dg = -ds$ in equation (4.2) to give

$$-g \frac{dZ}{dg} = p \qquad (4.4)$$

whence

$$Z = p \ln(m/g) = p \ln(1+s/g) \qquad (4.5)$$

(Searle and Sargent 1972).

The first prediction of the simple model, then, is that heavy-element abundance should be simply related to the net fraction of the mass of the system that has been changed from gas into stars. Observations of HII regions in Magellanic-type Irregular and blue

compact galaxies, together with estimates of total mass from rotation curves and HI mass from 21cm line emission, suggest that
this relationship does indeed hold within the uncertainties
(Lequeux et al 1979), taking the oxygen abundance as a measure of
Z ($Z \simeq 25\overline{(O/H)}$); see Figure 4. From this relation $p \simeq 0.004$
± 0.001.

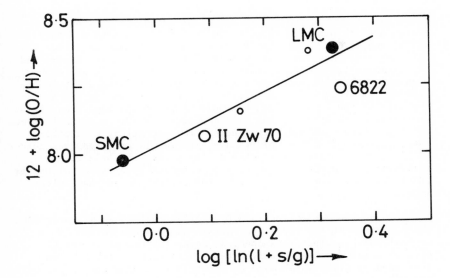

Fig. 4 Oxygen abundances in H II regions of Irregular and blue
        compact galaxies, as a function of the fractional mass of
        gas. The simple model predicts a line of unit slope, as
        shown.

A second consequence of the simple model is that abundances of
primary elements are proportional to their yields, and this con-
strains stellar evolution for a particular value of x, independently of ζ. Lequeux et al., following earlier work by the
Peimberts, derive a relationship between oxygen and helium abundance that can be expressed in the form

$$Y = Y_0 + Z(dY/dZ) \qquad\qquad (4.6)$$

with $Y_0 = 0.23$ from the Big Bang and $dY/dZ \simeq 3$ from stars. Now
the helium yield is virtually independent of whether mass loss
occurs in the earlier post main-sequence phase or not, since most
of the helium has already been produced by then anyway, but the
heavy-element yield can be reduced by a factor of 3 if the stars
lose nearly half their mass and $x \simeq 2$. The observed dY/dZ favours
substantial mass loss.

Some elements, notably s-process elements and some yet-to-be determined component of nitrogen, are secondary nucleosynthesis products for which the yield is proportional (other things being equal!) to the number of seed nuclei, previously produced by primary processes, i.e. to Z. In this case, instead of equation

(4.5), we have $\quad Z_2 \propto Z^2.$ $\hspace{4cm}$ (4.7)

Using data from HII regions in spiral galaxies, as well as Irregulars, it has been found by Smith (1975), Edmunds and Pagel (1978) and Alloin et al. (1979) that N and O abundances do not fit any such relationship. Rather we find that HII regions in Irr and Scd galaxies tend to have lower N/O than in our own Galaxy even when the oxygen abundance is the same, and that, when an abundance gradient is present in one galaxy, the variation in log N/O across the face of the galaxy is quite small. This means that the bulk of the nitrogen is primary, but with a yield relative to that of oxygen which varies by quite a large factor from one galaxy to another. The existence of sources of primary nitrogen is attested by a number of planetary nebulae, most notably one in the very N-deficient Irregular galaxy NGC 6822 where N is enhanced by a factor of nearly 200 relative to the HII regions of the system (Dufour and Talent 1980).

What do the variations in N/O signify? Alloin et al. attribute the main part of the effect to variations from one galaxy to another in the Salpeter exponent x, larger values of x giving greater weight to less massive stars which make N, relative to more massive stars which make O. Clearly it is possible to explain the effect in this manner, if one also adjusts ζ to get the yield right. Another effect is that of departures from instantaneous recycling. The age of the system, or more precisely the past star formation history, will affect the yield from stars of low mass and Edmunds and Pagel suggested that the nitrogen-deficient systems are "young" in the sense that star formation has been weighted to more recent times than in the solar neighbourhood. Both effects (as well as a contribution from secondary nitrogen in some circumstances) are probably involved.

Apart from He and N, we have data from extragalactic HII regions for Ne, S and Ar; within the errors, these vary in lockstep with oxygen as might be expected. Data for carbon, which will probably need ultra-violet observations, would be of considerable interest, as carbon may well receive a significant contribution from low-mass stars as well as being vulnerable to depletion by secondary nitrogen synthesis.

4.3 THE SIMPLE MODEL AND ITS FAILURE: THE SOLAR NEIGHBOURHOOD

van den Bergh (1962) and Schmidt (1963) first applied the Simple model to the solar neighbourhood and found it wanting.

The basic trouble lies in the distribution function of metallicities for long-lived stars, in particular the G dwarfs for which this function is known from U, B, V photometry; but similar considerations apply to M dwarfs too (Mould 1976). A simple formulation of the problem was given together with a survey of the data by Pagel and Patchett (1975) using the IR approximation, and the argument can be rephrased as follows.

$s \equiv \alpha\,S$ and $z \equiv Z/p$ are both monotonic functions of time increasing from $s = z = 0$ at $t = 0$ to $s = s_1 \simeq 0.9m$, $z = z_1$ at the present time $t = t_1$. Therefore s is an increasing function of z, observable now as the cumulative distribution of stars as a function of metallicity, and given from equations (4.3) and (4.5) by

$$s/m = 1 - g/m = 1 - e^{-z}. \qquad (4.8)$$

The corresponding differential distribution is

$$m^{-1}\frac{ds}{dz} = e^{-z} \qquad (4.9)$$

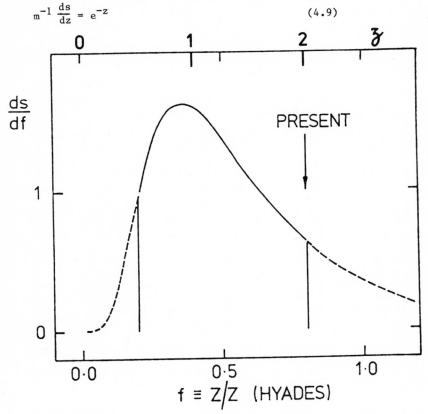

Fig 5. Differential distribution of G dwarfs as a function of mean metallicity relative to the Hyades.

whereas the observed distribution (corrected for observational errors) is a lognormal curve with somewhat arbitrary truncations (Fig 5) with a maximum somewhere around $z \simeq 1$. In other words the observed distribution is much narrower than the Simple model predicts, with many fewer stars near $z = 0$. (The model predicts no stars actually having $z = 0$, so that the failure to observe such stars is no anomaly in itself; but the halo stars occurring to the left of the lower truncation point are too few in number to affect Fig 5 significantly.)

Various more or less ad hoc assumptions have been introduced to patch up this disagreement:-

(i)   Variable IMF with greater yields in the past (Schmidt 1963). In its extreme form, termed "prompt initial enrichment" (PIE), this leads to the same distribution but with an initial abundance $z_0 \simeq 0.3z_1$ and fits the data very well at the expense of an extra parameter. Truran and Cameron (1971) suggested PIE from an early generation consisting of massive stars alone.

(ii)   PIE from massive stars in the halo, before star formation in the disk got under way (Ostriker and Thuan 1975). This gives a still better fit (as it allows for the few halo stars observed) and is probably quite relevant; a contribution from this effect is included in dynamical collapse models (Larson 1976; Tinsley and Larson 1978).

(iii)   Metal-enhanced star formation (Talbot and Arnett 1973, 1975; Talbot 1974). An early extreme form of MESF is mathematically equivalent to PIE and therefore fits the data equally well. Later more realistic versions are less drastic and insufficient to solve the problem by themselves.

(iv)   Inflow of unprocessed material. A steady inflow (Oort 1970) can cut down, neutralise or even reverse the normal enrichment as a function of time, reaching in the end a steady state with $z = 1$ (Larson 1972). This may be relevant to the age-metallicity relation, which shows some signs of going flatter during the last 3×10^9 years (Twarog 1979) and to the abundances of light elements (Reeves and Meyer 1978). A steady inflow, however, does not by itself give a good fit to the number-metallicity relation (Pagel and Patchett 1975). A decaying inflow, on the other hand, such as is associated with dynamical collapse models of galaxy formation (Larson 1976) is quite capable of clearing up the problem (Lynden-Bell 1975), or explaining PIE if one likes to put it that way, but at this point we must abandon the simple model (for the time being at least) and use a more general formulation.

CHEMICAL EVOLUTION OF NORMAL GALAXIES

Before doing so, I mention parenthetically that the halo stars
and globular clusters, taken by themselves, do have a number-
metallicity distribution that fits the simple model, but with a
much lower yield. Various interpretations of this result have been
given by Hartwick (1976), Searle (1977) and Searle and Zinn (1978).

4.4 DYNAMICAL COLLAPSE MODELS

From the ages of globular and galactic clusters (Demarque and
McClure 1977) it appears that the formation of the Galactic disk
was a long drawn-out process as in the dynamical collapse models
of Larson (1976) which make specific predictions about star formation rates and chemical evolution and imply that the simple model
is physically unrealistic for galaxies like our own. In these
models the "G dwarf" problem is solved by a combination of two
factors: (i) the Ostriker-Thuan effect already mentioned; and (ii)
concentration of star formation in a high-density region into
which unprocessed gas is flowing in from the outside at least for
a certain time. This type of model has been explored further by
Vader and de Jong (1979), Chiosi (1980) and Chiosi and Matteucci
(1980), who consider the effect of vertical inflow along cylindrical shells perpendicular to the galactic plane, and an analytical treatment of its chemical evolutionary consequences has been
given by Lynden-Bell (1975).

We can characterise these models in a simple, very crude way
as follows. Basically there are two phases. In phase (i), unprocessed gas enters the region of star formation at a rate that is
initially more than sufficient to balance the rate of star formation. In this case we use equation (4.2) with $Z_f = 0$, i.e.

$$g \frac{dz}{ds} = 1-z \frac{dm}{ds} \qquad (4.10)$$

As stars are formed and z goes up, $\frac{dz}{ds}$ will go down and if a

steady state is reached it will have

$$Z = p/ \frac{dm}{ds} \qquad (4.11)$$

which can be greater, less than or equal to p according to
whether g is decreasing, increasing or constant. In this phase
all stars have the same Z and it may go on for quite a long time,
giving an equivalent of prompt initial enrichment. In phase
(ii), formation of the disk is completed and we run out of
material to accrete, thus returning to something like the Simple
model. A third stage may be envisaged corresponding to steady
inflow at a much lower rate than in phase (i). In this case the
enrichment predicted by the simple model will flatten off and
eventually go down until, after very long times, Z is down to p
again (Pagel and Patchett 1975).

4.5 SKETCH OF A SELF-CONSISTENT MODEL FOR THE SOLAR NEIGHBOURHOOD

We have given the stellar metallicity distribution $s(z)$ and the age-metallicity relation. These can be put together to trace the past rate of star formation using

$$\frac{ds}{dt} = \frac{ds}{dz}\frac{dz}{dt}. \tag{4.12}$$

A particularly simple situation results if we assume $\frac{dz}{dt}$ = const.,

which fits either Mayor's or Twarog's data over quite a wide range, though not the whole range in the case of Twarog's date. For the present preliminary exploration, we shall ignore this complication. We then have the very simple situation that Fig 5 represents not only the differential metallicity distribution, but also the stellar birthrate as a function of time since the metallicity was z_0. The curve will be seen to bear a family relationship to the star formation rates computed by Larson (1976).

Assume that, in phase (i), as a result of inflow, the metallicity had some steady value z_0 corresponding to equation (4.11) while the total mass and gas mass increased from zero to m_0 and g_0 respectively, whereafter we have phase (ii) with no further inflow, until $g = g_1$ etc at the present time in accordance with the Simple Model. m_0 is thus the present total mass of the system: about 100 $M_\odot$ pc^{-2} for the solar cylinder.

Unknown parameters in our modified simple model are the relative amounts of gas and stars accumulated during the steady-state phase (i) and the amount of gas used up since. As constraints we have

$$z_0 = \frac{ds}{dm} = \frac{1-dg}{dm} \approx 1-\frac{g_0}{m_0} \tag{4.13}$$

$$z_1 = z_0 + \ln(g_0/g_1) \tag{4.14}$$

and (from observation)

$$z_1 \approx 4z_0. \tag{4.15}$$

Our final constraint comes from equation (4.9): z_1 (and hence the yield) has to have such a value that the right part of the curve in Fig 5 is an exponential in $z - z_0$. This leads to

$$z_1 \approx 2$$

corresponding to a yield of about 0.01, somewhat larger than was
found for HII regions in Irregular galaxies (but not by a factor
exceeding the discrepancy between the Sun and Orion!). The other
parameters then are $z_0 = 0.5$, $g_0 = 0.5m_0$, $s_0 = 0.5m_0$, $g_1 = 0.09m_0$,
all of which seem reasonable. Thus the dynamical collapse picture
fits all the data for the solar neighbourhood pretty well.

Finally, we note that the maximum rate of star formation, corresponding to about 5 times the present mass of gas, is about 3 times
the present rate, a result that is not incompatible with Schmidt's
law for n = 1 (cf. Vader and de Jong 1979) when minor fluctuations
are allowed for.

5. ABUNDANCE GRADIENTS AND MASS-METALLICITY RELATION IN GALAXIES
    The gradients observed in elliptical galaxies support the
dissipative collapse model of galaxy formation (Larson 1974) in
which enriched gas loses its energy and systematically falls into
the central regions, piling up metals there. This effect can be
derived from equation (4.2) if we imagine a region with partially
processed gas falling in, in which a steady state is approached,
when we have (Tinsley 1980)

$$Z \rightarrow Z_f + p / \frac{dm}{ds} \qquad (5.1)$$

which may occasionally produce a situation with $Z \gg p$.

However, the metallicity in a stellar population, averaged over
the entire galaxy, is unlikely to differ much from p, so that
differences between entire galaxies suggest a difference in the
effective yields, perhaps due to loss of gas in galactic winds or
mergers, an effect which can also be related to the paucity of gas
in Ellipticals and the presence of iron in intergalactic gas
in rich clusters of galaxies (Tinsley and Larson 1979; Hartwick
1980; Tinsley 1980).

Why are there radial abundance gradients in Spirals? Searle
and Sargent (1972), applying the Simple Model to separate cylindrical zones at right angles to the galactic plane, suggested
attributing them to increased efficiency of star formation in the
interior regions, symptomised by an inwardly decreasing ratio of
surface density of gas to total surface density. Thus, if the
surface densities decrease outwards exponentially with scale
lengths R_G and R_T respectively, then equation (4.5) predicts

$$z = \text{const} \ -R \ (\frac{1}{R_T} - \frac{1}{R_G}), \qquad (5.2)$$

i.e. a linear decrease in z with radial distance R which fits the data in the solar neighbourhood (8 < R < 14kpc) quite well (Pagel 1979b). Despite the various elaborations in galactic evolutionary models that have taken place since then, the same basic idea underlies most theoretical treatments of this effect, even in the case of the dynamical collapse models although here other effects are involved as well.

There are several reasons for doubting whether this explanation is wholly adequate. For one thing, there seems to be little difference between R_T and R_G in our Galaxy when H_2 molecules are taken into account (Gordon and Burton 1976), although the H_2 density deduced from CO needs to be corrected for any abundance gradient of carbon in the gas phase (cf. Talbot 1980). Even without H_2, however, there is rather little difference between R_T and R_G in some galaxies (Bosma 1978). Thirdly, owing to departures from the IR approximation, enrichment in the Simple Model comes to an end at $z \simeq 3$ or so, which severely limits the amount of differential enrichment that one can get in such models between here and the Galactic centre, and fourthly the more realistic dynamical collapse models tend to give a smaller radial gradient than the Simple model, because of the effect of infall (Tinsley and Larson 1978).

We do not know for certain how strong a gradient we actually have between here and the Galactic centre, or in corresponding regions of other galaxies with HII regions, partly because of the discrepancy between the Sun and Orion and partly because of having to use photoionisation models for the low-excitation, high-abundance HII regions near the centres of large galaxies, but a reasonable guess is a factor of 3. This is somewhat higher than dynamical collapse models predict, although it has to be said that equation (5.1) actually allows considerable latitude. The gradient can perhaps be enhanced by moderate radial inflows $\sim 1 kms^{-1}$ or $\frac{1}{2}M_\odot yr^{-1}$ (Mayor 1979), caused by accretion of halo or intergalactic material of low angular momentum, either in the past or in the present, leading to enhanced abundance gradients in normal and intermediate-barred spirals. In Irregulars and parts of barred spirals, on the other hand, strong non-circular motions of gas are either observed or predicted (cf. Roberts et al. 1979), and these could perhaps neutralise any tendency to build up a gradient.

6. CONCLUSION

In the present state of uncertainty about many problems, the theory of chemical evolution of galaxies sometimes appears as a rather flabby component of a larger jig-saw puzzle which is open to the charge of explaining (nearly) everything and predicting nothing. It is actually not quite as bad as that, since both Irregular galaxies and the solar neighbourhood have yielded to models that are plausible in themselves with relatively few ad hoc

assumptions, and some quite specific constraints have been placed
on stellar evolutionary models, but some worries remain. Two
problems in which we may see further interesting developments
stimulated by chemical arguments are (i) inflows or other
dynamical effects suggested by the existence of abundance
gradients in some disk-like galaxies but their weakness or
absence in others; and (ii) possible differences in the effective
ages and IMF's of stellar populations in different galaxies.
The second question is one of many that may be resolved with the
aid of the Space Telescope in the next few years.

B. E. J. PAGEL

REFERENCES

Aitken, D. K., Roche, P. F., Spenser, P. M., and Jones, B. 1979,
    Ap. J., 233, 925

Alloin, D., Collin-Souffrin, S., Joly, M., and Vigroux, L. 1979
    Astr. Ap. 78, 200

Audouze, J., and Tinsley, B. M. 1976, Ann. Rev. Astr. Ap., 14, 43

Baschek, B. 1979, in Les Elements et leurs Isotopes dans l'Univers,
    A. Boury, N. Grevesse and L. Remy Battiau (eds), Liège, p.327

Bosma, A. 1978, Thesis, Groningen

Burbidge, E. M., Burbidge, G. R., Fowler, W. A., and Hoyle, F.
    1957, Rev. Mod. Phys., 29, 547

Chiosi, C. 1980, Astr. Ap., 83, 206

Chiosi, C., and Caimmi, R. 1979, Astr. Ap., 80, 234

Chiosi, C., and Matteucci, F. 1980, preprint

Demarque, P., and McClure, R. D. 1977, in The Evolution of
    Galaxies and Stellar Populations, B. M. Tinsley and R. B.
    Larson (eds), Yale University Observatory, New Haven, p.199

Dopita, M. A., D'Odorico, S., and Benvenuti, P. 1980, Ap. J.,
    236, 628

Dufour, R. J., Talbot, R. J., Jensen, E. B., and Shields, G.

    1980, Ap. J., 236, 119

Dufour, R. J., and Talent, D. L. 1980, Ap. J., 235, 22

Edmunds, M. G., and Pagel, B. E. J. 1978, Mon. Not. Roy. Ast.
    Soc., 185, 77P

Edmunds, M. G., and Pagel, B. E. J. 1980, in preparation

Eggen, O. J., Lynden-Bell, D., and Sandage, A. R. 1962, Ap. J.,
    136, 748

Faber, S. M. 1977, in The Evolution of Galaxies and Stellar
    Populations, B. M. Tinsley and R. B. Larson (eds), New
    Haven, p. 157

Gordon, M. A., and Burton, W. B. 1976, Ap. J., 208, 346

234

Harris, W. E., and Canterna, R. 1979, Ap. J. Let., 231, L19

Hartwick, F. D. A. 1976, Ap. J., 209, 418

Hartwick, F. D. A. 1980, Ap. J., 236, 754

Janes, K. A. 1979, Ap. J. Suppl. 39, 135

Jensen, E. B., Strom, K. M., and Strom, S. E. 1976, Ap. J.
    209, 748

Larson, R. B. 1972, Nature Phys. Sci., 236, 7

Larson, R. B. 1974, Mon. Not. Roy. Ast. Soc., 166, 585

Larson, R. B. 1976, Mon. Not. Roy. Ast. Soc., 176, 31

Larson, R. B. 1977, in The Evolution of Galaxies and Stellar
    Populations, B. M. Tinsley and R. B. Larson (eds),
    New Haven, p. 97

Lequeux, J., Peimbert, M., Rayo, J. F., Serrano, A., and
    Torres-Peimbert, S. 1979, Astr. Ap. 80, 155

Lynden-Bell, D. 1975, Vistas in Astr., 19, 299

Mayor, M. 1976, Astr. Ap., 48, 301

Mayor, M. 1979, Mem. Soc. Astr. Italiana, 50, 157

Mayor, M., and Martinet, L. 1977, Astr. Ap., 55, 227

Mezger, P. G., Pankonin, V., Schmidt-Burgk, J., Thum, C., and
    Wink, J. 1979, Astr. Astrophys., 80, L3

Miller, G. E., and Scalo, J. M. 1979, Ap. J. Suppl., 41, 513

Mould, J. R. 1976, Mon. Not. Roy. Ast. Soc. 177, 47P

Oort, J. H. 1970, Astr. Astrophys., 7, 381

Ostriker, J. P., and Thuan, T. X. 1975, Ap. J., 202, 353

Pagel, B. E. J. 1979a, in Les Elements et leurs Isotopes dans
    l'Univers, A. Boury, N. Grevesse and L. Remy Battiau (eds),
    Liège, p. 261

Pagel, B. E. J. 1979b, in Stars and Star Systems,
    B. E. Westerlund (ed), Reidel, p.17

Pagel, B. E. J., and Patchett, B. E. 1975, M.N.R.A.S, 172, 13

Pagel, B. E. J., Edmunds, M. G., Fosbury, R. A. E., and
     Webster, B. L. 1978, Mon. Not. Roy. Ast. Soc., 184, 569

Pagel, B. E. J., Edmunds, M. G., Blackwell, D. E., Chun, M. S.,
     and Smith, G. 1979, Mon. Not. Roy. Ast. Soc., 189, 95

Pagel, B. E. J., Edmunds, M. G., and Smith, G. 1980,
     Mon. Not. Roy. Ast. Soc., in press

Peimbert, M. 1978, in Planetary Nebulae: Observations
     and Theory, I.A.U. Symposium no. 76, Y. Terzian (ed),
     Dordrecht: Reidel, p. 215

Peimbert, M. 1979, in The Large-scale Characteristics of the
     Galaxy, I.A.U. Symposium no. 84, W. B. Burton (ed),
     Dordrecht: Reidel, p.79

Reeves, H., and Meyer, J.-P. 1978, Ap. J., 226, 613

Roberts, W. W., Jr., Huntley, J. M., and van Albada, G. D.
     1979, Ap. J., 233, 67

Salpeter, E. E. 1955, Ap. J., 121, 161

Sargent, W. L. W., Young, P. J., Boksenberg, A., and
     Tytler, D. 1980, Ap. J. Suppl., 42, 41

Schmidt, M. 1959, Ap. J., 129, 243

Schmidt, M. 1963, Ap. J., 137, 758

Searle, L. 1971, Ap. J., 168, 327

Searle, L. 1977, in The Evolution of Galaxies and Stellar
     Populations, B. M. Tinsley and R. B. Larson (eds), Yale
     University Observatory, p. 219

Searle, L., and Sargent, W. L. W. 1972, Ap. J., 173, 25

Searle, L., and Zinn, R. 1978, Ap. J., 225, 357

Shaver, P. A., McGee, R. X., and Pottasch, S. R. 1979,
     Nature, 280, 476

Shields, G. A., and Searle, L. 1978, Ap. J., 222, 821

Shields, G. A., and Tinsley, B. M. 1976, Ap. J., 203, 66

Smith, H. E. 1975, Ap. J., 199, 591

Talbot, R. J. 1974, Ap. J., 189, 209

Talbot, R. J. 1980, Ap. J., 235, 821

Talbot, R. J., and Arnett, W. D. 1971, Ap. J., 170, 409

Talbot, R. J., and Arnett, W. D. 1973, Ap. J., 186, 60

Talbot, R. J., and Arnett, W. D. 1975, Ap. J., 197, 551

Talent, D. L., and Dufour, R. J. 1979, Ap. J., 233, 888

Terlevich, R., Davies, R. L., Faber, S. M., and Burstein, D.
    1980, Mon. Not. Roy. Ast. Soc., in press

Tinsley, B. M. 1980, submitted to Fundamentals
    of Cosmic Physics, 5, 287

Tinsley, B. M., and Larson, R. B. 1978, Ap. J., 221, 554

Tinsley, B. M., and Larson, R. B. 1979, Mon. Not. Roy.
    Ast. Soc,, 186, 503

Tomkin, J., and Lambert, D. L. 1980, preprint

Trimble, V. 1975, Rev. Mod. Phys., 47, 877

Truran, J. W., and Cameron, A. G. W. 1971, Ap. Sp. Sci.,
    14, 179

Twarog, B. A. 1979, Thesis, Yale University

Vader, J. P., and de Jong, T. 1979, in Les Elements
    et leurs Isotopes dans l'Univers, A. Boury,
    N. Grevesse and L. Remy Battiau (eds), Liège, p.529

van den Bergh, S. 1962, A. J., 67, 486

Ward, M., Penston, M.V., Blades, J.C., and Turtle, A.J. 1980
    preprint

Wilson. A.S., Ward, M., and Blades, J.C. 1980, in preparation.

THE RATE OF STAR FORMATION IN GALAXIES

Barry F. Madore

David Dunlap Observatory
Department of Astronomy
University of Toronto

1 INTRODUCTION : MOTIVATION AND CONSTRAINTS

Investigations into the rate of star formation in external
galaxies are, in most cases, doubly motivated. For some, the global
aspects of star formation are of most interest, for others it is the
localized event that is the object of inquiry. Those studying the
evolution of galaxies have long sought simple parametric relations
for the average rate at which gas is processed into and through
stars, enriched in metals and returned to the interstellar medium.
Those interested in star formation, as a physical process in itself,
looked to the extragalactic evidence on the rate of star formation
as providing a wide range of environments constraining their theory
or motivating new directions of thought. Both of these approaches
are still very much in their infancy and show all of the side effects
of idealistic optimism followed by the inevitable frustrations.
Nevertheless, the study of star formation promises to continue to
be an active point of intersection between theory and observation
and for this reason alone it deserves close monitoring.

Excellent modern reviews of the theory and observations bearing
on the rate of star formation are provided from a theorist's point
of view by Larson (1977), Talbot (1977) and Silk (1980), while from
an observer's perspective by Kerr (1977), Mezger & Smith (1977) and
Lequeux (1980). The complex inter-relation between determining
the rate, in concert with other aspects of star formation such as
the mass function, are treated in the well-balanced critical summa-
ries by Scalo (1978) and Miller & Scalo (1979). Many ambiguities
in definitions and methodology are pointed out in these discussions;
they are worth noting. The following will be a topical review of
the many types of star formation mechanisms suggested to date but
set in a speculative extragalactic context.

2 THE RATE OF STAR FORMATION : A DEFINITION

Aside from isolated suggestions to the contrary (eg. Ambartsumian, 1955) stars are thought to have condensed out of the interstellar (sic) medium. The time derivative of this conversion of

mass, M_g, from a tenuous state into a self-luminous/condensed stellar phase, M_*, is the rate of star formation,

$$f = + \frac{dM_*}{dt} = - \frac{dM_g}{dt} . \tag{1}$$

As simple as this definition is, it has been much abused in the transition to observational status.

2.1 Quiescent Star Formation

2.1.1 Schmidt's Law : A Parameterization and A Critique

Most dynamical timescales vary as the inverse square root of the mean density of the system. This generalization appears to hold true for pendula, variable stars and the mechanical fate of the Universe. The <u>initial</u> gas density ρ_g° therefore might seem to be a natural primary parameter to investigate in any discussion of the <u>rate</u> of star formation. One, amazingly popular parameterization of this form,

$$f \sim (\rho_g^\circ)^n \tag{2}$$

is known as Schmidt's Law. In a series of papers and discussions Schmidt (1959, 1962, 1963) used this expression to study the time evolution of star formation in our Galaxy. His final conclusion was that $n \sim 1-2$.

One of the most graphic illustrations of evidence in favour of $n > 1$ is found in the oft-cited comparison of the galactic plane scale heights for the gas with respect to the young stars. As is known the <u>neutral</u> hydrogen is much more broadly distributed out of the plane of the Galaxy than are the young stars that have recently formed from it. <u>If</u> the star formation rate, as represented by the present star density, is to follow Schmidt's Law then regions of high gas density must produce more stars than regions of low gas density; that is n must be greater than unity. As appealing as this simple demonstration is, it is filled with precipitous pitfalls which apply to many of the studies which followed Schmidt:

Figure 1. The distribution of the brightest stars in M101 as taken from an ultra-violet exposure using the 3.6 m Canada-France-Hawaii Telescope. The distribution is almost identical to the 21 cm intensity map of Allen & Goss (1979).

(1)   There is no reason to believe that any <u>instantaneous</u> measurements of the <u>numbers</u> of stars (or their densities) is proportional to the <u>rate</u> of their production. A region with stars now is not necessarily an eternally active region, but may simply be caught in the act after aeons of a slow approach.

(2)   Large spatial averages are likewise not equivalent to temporal averages (Madore 1977). Due to the finite lifetimes of bright stars regions of high gas density are always over-represented with respect to regions of low gas density.

(3)   In the presence of severe or varying amounts of depletion, spatial averages over regions that both have and have not undergone recent star formation will inject observational noise until eventually an <u>anticorrection</u> of observed star densities with remaining gas densities will result. (The fact that strong correlations are found indicates that the efficiency of star formation is probably quite low (Blaauw, 1964; Madore, 1977)).

(4)   Gas densities are not necessarily represented by the neutral hydrogen observations. In our Galaxy it is thought (Gordon & Burton, 1976) that a significant fraction of the total gas content is in molecular form. If this component is more restricted in scale height, as seems to be the case, then the evidence for n > 2 begins to vanish. One could, of course, argue that the molecular clouds are sequential in the star formation process, that their formation rate varies as some high exponent of the neutral gas density and that star formation proceeds more linearly with the molecular component.

A much more complex situation must exist. The gaseous disk is both dissipative and tends to cool and contract but it is also responsive to star formation and tends to heat and re-inflate. Feed-back processes may therefore be very important in regulating star formation. (A powerful demonstration of these effects is found in the film described by Miller & Smith (1979) in which a collapsing three-dimensional galaxy self-destructs in its first free-fall time). Only one other attempt to model this situation, for disk galaxies, is found in Talbot & Arnett (1975). Certainly more attention needs to be focused on the detailed structure of the interstellar medium, not only on how it produces stars but how

Figure 2. The distribution of brightest stars in M81 as taken from an ultraviolet exposure using the 3.6 m Canada-France-Hawaii Telescope. Note the central minimum and the partially dissolved ring which is also visible in the 21 cm intensity maps of Rots & Shane (1975).

it, in turn, responds to star formation. The amazing maps of Colomb, Pöppel & Heiles (1980) should give serious pause to all who neglect the fine structure of our Galaxy.

(5) While the molecular component may not be distributed out of the plane in the same way as the neutral component there are some interpretations of the galactic CO data that it is not distributed radially from the galactic centre in the same way as the neutral hydrogen (Gordon & Burton, 1976). Freedman & Madore (1980) have suggested that accounting for metallicity gradients can in fact restore a constant ratio and that the extragalactic evidence on the correlations between bright stars (Figures 1, 2, 3, & 4) and HI maps argue that this is the case in other galaxies as well.

The rate of star formation is certainly not just a function of the one parameter, gas density, but this naive first step can lead to a better understanding of the processes which all contribute to galaxy evolution. This realization is on-going.

2.2 Stimulated Star Formation

2.2.1 Short-Range Modes

The possibility of a form of star formation that is not self-initiating was first put forward by Opik (1953) who suggested that supernovae might trigger subsequent events. This mechanism has recently found considerable observational support and is reviewed by Herbst & Assousa (1978). Oort's (1954) suggestion that ionization fronts, driven by a previous generation of O stars and giving rise to pressure-induced star formation has been developed most actively of late by Elmegreen & Lada (1977). In addition stellar winds from early-type stars have also been suggested (Blitz, 1978) as yet another means of producing the external pressure needed for star formation. An essential aspect of each of these mechanisms is that all are highly localized phenomena depending very critically on the pre-existing configuration of stars and gas on scales of less than 100 pc. or so (Lada, Blitz & Elmegreen, 1978).

Figure 3. The distribution of brightest stars in NGC 300, as taken from an ultraviolet exposure using the CTIO 4 meter reflector. Note that the spiral structure is even more ragged than that of the simulations in Fig. 5.

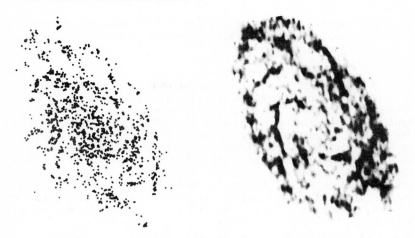

Figure 4. The bright star distribution (left) in M33 (Madore, 1978) in comparison to the neutral hydrogen distribution (Newton, 1980). Many features repeat in both maps with exacting detail.

If galaxy evolution, via the general rate of star formation, is dominated by such processes acting on small scales, it may require the modelling of some 10^9 evolving and interacting "cells" to specify the problem fully! An approach to this limit is in fact occurring asymptotically. Mueller & Arnett (1976) began simulating the ragged two-dimensional spiral structure of galaxies using deterministic star formation algorithms acting on cells of about a half a kiloparsec each, set in a differentially rotating disk. Later Gerola & Seiden (1978), Seiden, Schulman & Gerola (1979) and Seiden & Gerola (1979) introduced a probablistic element into the star formation rate, decreased the cell size to about 200 parsecs and investigated a wide range of other parameters necessary to define the models. Time evolving simulations incorporating stellar evolution, more realistic gas distributions, stellar mass loss and gas depletion by stars have been completed by Freedman (1980). One of her galaxies is shown in Figure 5.

In these two-dimensional visual models of galaxies the rate of star formation has become a "semi-empirical" function of a number of parameters, obliquely related to the physical world. The most important parameters are the probability of stimulated star formation, the probability of spontaneous star formation, a quiescent period, the rate of shearing of adjacent cells and the timestep, determined essentially by the cell size. Primarily, the appearance of coherent spiral structure and its stability are used to constrain the wide range of available parameters and so it seems that one of the biggest stumbling blocks of early spiral structure theories, "the wind-up problem" is now a criterion for success in modern studies. Hopefully this time the underlying "physics" of these new models has a realistic basis and can be eventually interpreted.

2.2.2 Long-Range Modes

Of course, differentially sheared associations of bright stars (Figures 3 & 4) are not <u>the</u> motivating observations for all spiral structure theories. The so-called "grand-design" two-armed spirals typified (in the extreme!) by such objects as M51 and NGC 4622 are the subject of much more sophisticated discussions. The density-wave theory of Lin & Shu (1964), in which a gravity-induced spiral of enhanced potential rotates rigidly within the disk of a galaxy, can produce large-scale shocks in the interstellar medium (Fugimoto, 1966; Roberts, 1969). If sufficient compression is induced (depending on the local sound speed in the gas and the relative motion of the density wave with respect to the ambient medium) star formation may result. The list of parameters, regulating the rate at which this star formation mechanism can act, includes the pattern speed, the inclination of the spiral to the direction of propagation, the amplitude of the potential well, the galactic rotation curve, the local thermal conditions of the gas and so on.

Unfortunately the density waves are neither self-generating, nor self-sustaining (Toomre, 1969). Kormendy & Norman (1979) have investigated the <u>quality</u> of spiral structure that galaxies develop in the hope of discovering the mechanism for inducing and/or maintaining spiral structure. In the absence of a central bar or a large (presumably interacting) companion, galaxies with strong differential rotation show no well-defined grand-design spiral pattern. Of the remainder, only those galaxies or parts of galaxies with fairly rigid rotation curves appear to be able to maintain grand-design spirals on their own. Even for those galaxies with density waves, the effect on the rate of star formation may sensitively depend on radial gas flow which is usually ignored and certainly not well understood (Cassé, Kunth & Scalo, 1979).

The message so far seems clear: both one-zone and one-parameter models for galaxy evolution are bound to fail. Both the internal dynamics of the galaxy and its dynamical environment have a profound

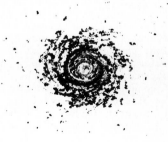

Figure 5. One frame of a galaxy evolution simulation (Freedman, 1980) using realistic rotation curves, gas density distributions, stellar evolution, and mass return.

influence on the manner in which star formation proceeds and they must be taken into account. There is more data to this effect.

2.2.3 Very Long-Range Modes : Quiescent to Cataclysmic

Galaxies interact. While even a casual inspection of the Atlas of Peculiar Galaxies (Arp 1966) is convincing evidence that the galaxies respond dramatically to one another, all but a few of the parameters relating to interaction have in any way been quantified. What is the response time of a galaxy to a collision? Are there critical interaction distances or mass ratios? What are the past and present fequencies of these interactions? More broad, statistical (Madore 1980; Arp & Madore, 1977) and detailed, individual studies (eg. Sandage, 1963) are needed.

The existing data show already that such studies will be very productive. For instance, it has been known for some time (Page 1975; Noerdlinger 1979) that galaxies of similar Hubble type tend to be found in binary systems. Recent evidence (White & Valdes, 1980) indicates that these galaxies are also systematically brighter than the equivalent galaxies in the field. Rough colours show that these galaxies are also bluer (Sharp & Jones, 1980) indicating that star formation is proceeding more rapidly in binary galaxies than in isolated systems. A new analysis (Madore 1981) of the photoelectric colour data of Tomov (1978) makes it clear that the interaction, rather than the initial conditions are dominant: galaxies with low apparent separations are bluer than those with large separations out to a cut-off radius of about 60 kiloparsecs. The gas deficiency of spirals in the Virgo cluster (Davies & Lewis, 1973) and the correlations of separation versus morphology of companions found by Chernin, Einasto & Saar (1976) and possibly even the S0 phenomenon may be related to enhanced star formation due to environment, rather than its termination due to stripping.

Some quantitative analysis is, in fact, already available. For instance, Larson and Tinsley (1978) have estimated that the UBV colours of peculiar/interacting galaxies are sufficiently scattered away from the colours of normal galaxies that they can be interpreted as being due to induced bursts of star formation lasting as little as 2×10^7 years and involving up to five percent of the total mass. The photoelectric Hα studies being presently undertaken by Kennicutt (private communication) should greatly extend these conclusions.

Galaxies collide. Perhaps the most graphic illustrations of this are found in the ring galaxies. These systems are most likely (Chatterjee, 1979) the result of a "gravitational splash" caused by the direct hit of a massive companion travelling through the plane of a gas-rich spiral. The differential response (Lynds & Toomre, 1976) of the disk to the passage of such a companion results in an expanding ring of star formation, evidenced by the ionized gas and bright star distribution (Figure 6) which give these systems their name.

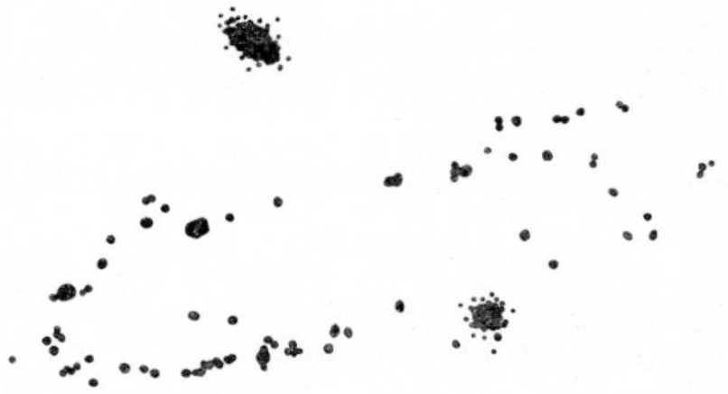

Figure 6 (left). The distribution of HII regions in the ring galaxy AM 064-741 and the relative positions of two companion ellipticals thought to be responsible for producing the ring.

Figure 7 (right). The distribution of HII regions (Arp & Brueckel, 1973) in M31 and the relative positions of M32 and NGC 205 which may be responsible for stimulating star formation in the disk of M31.

The dynamical models of Toomre (1978) and Few (1979) treat only the stellar response and indicate that a companion of comparable mass is necessary to perturb the entire disk. However, in analogy with the density-wave theory, where the gas response is non-linear and is expected to initiate star formation with only a change of a few percent in the potential, it is reasonable to expect that even low-mass companions passing through the disk could be responsible for stimulating star formation. A case in point is M31, where a ring of HII regions (Figure 7) and a centrally depressed gas distribution may be due to the repeated passage of the tidally-stripped companion, M32. The stellar distribution in M81 (Figure 2) may also be the remnant of a collision with one of its companions NGC 3077.

If large companions with plunging orbits can stimulate global star formation, what is the cut-off? Perhaps the effects continue down to very low masses, in which case even the size of the globular cluster population of a galaxy may play a role in modifying the rate of star formation. This would occur not in one or two spectacular bursts but through a continual hailstorm of events. How these effects are to be modelled and how they were different in the early history of the Universe remains to be seen.

3 CONCLUSIONS

Galaxies of bewildering variety appear capable of producing stars, in isolation, in pairs and in clusters. The rate at which this process occurs and how long it is allowed to continue now appears to be a much broader function of local gas properties, internal galactic dynamics and global environment than was once realized and investigations of a synoptic nature are now needed. Galaxies are aware of their environment and respond to it; theory and observations now need to take this more fully into account.

Allen, R.J. & Goss, W.M. (1979). The Giant Spiral Galaxy M101 : V. A Complete Synthesis of the Distribution and Motions of the Neutral Hydrogen. Astron. Astrophys. Suppl., 36, 135-62.

Ambartsumian, V.A. (1955). Stellar Systems of Positive Total Energy. Observatory, 75, 72-8.

Arp, H.C. (1966). Atlas of Peculiar Galaxies, California Institute of Technology, Pasadena = Ap.J. Suppl., 14, 1-20.

Arp, H.C. & Brueckel, F. (1973). Diameters of HII Regions in M31 and Comparison with the Largest Regions in M33. Ap.J., 179, 445-52.

Arp. H.C. & Madore, B.F. (1977). Preliminary Results from the Catalogue of Southern Peculiar Galaxies and Associations. Q.J.R.A.S., 18, 234-41.

Baldwin, J.E. (1976). Gas in Galaxies of the Local Group. In The Galaxy and the Local Group. R.O.G. Bull. No. 182, ed. R.J. Dickens & J.E. Perry, 207-14, Royal Greenwich Observatory, Herstmonceux.

Blaauw, A. (1964). The O Associations in the Solar Neighborhood. Ann. Rev. Astr. Astrophys., 2, 213-46.

Blitz, L. (1978). Star Forming Clouds Towards the Galactic Anti-Centre. In Proceedings of the Gregynog Workshop on Giant Molecular Clouds, ed. P.M. Solomon & M.G. Edmunds, 211-30, Pergamon, Oxford.

Cassé, M., Kunth, D. & Scalo, J.M. (1979). A Constraint on the Influence of Density Waves on the Rate of Star Formation. Astron. Astrophys., 76, 346-9.

Chatterjee, T.K. (1979). Ring Galaxies - A Review. Bull. Astr. Soc. India, 7, 32-7.

Chernin, A., Einasto, J. & Saar, E. (1976). The Role of Diffuse Matter in Galactic Coronas. Astrophys. Space Sci., 39, 41-64.

Colomb, F.R., Pöppel, W.G.L. & Heiles, C. (1980). Galactic HI at $|b| \geq 10°$. II. Photographic Presentation of the Combined Southern and Northern Data. Astron. Astrophys. Suppl., 40, 47-56.

Davies, R.D. & Lewis, B.M. (1973). Neutral Hydrogen in Virgo Cluster Galaxies. M.N.R.A.S., 165, 231-44.

Elmegreen, B.G. & Lada, C.J. (1977). Sequential Formation of Subgroups in OB Associations. Ap.J., 214, 725-41.

Few, J.M.A. (1979). The Dynamics and Morphology of Ring Galaxies. Unpublished Doctoral Thesis, Cambridge University.

Freedman, W.L. (1980). Deterministic Models of Galaxy Evolution. Masters Thesis, University of Toronto.

Freedman, W.L. & Madore, B.F. (1980). Concerning the Radial Distribution of Gas in the Galaxy. (preprint).

Fugimoto, M. (1966). Gas, Instabilities Across Model Spiral Arms. In Non-Stable Phenomena in Galaxies, IAU Symp. No. 29., Academy of Sciences of Armenia, SSR, Yerevan, (in Russian) 453-63.

Gerola, H. & Seiden, P.E. (1978). Stochastic Star Formation and Spiral Structure of Galaxies. Ap.J., 223, 129-39.

Gordon, M.A. & Burton, W.B. (1976). Carbon Monoxide in the Galaxy I. The Radial Distribution of CO, H_2 and Nucleons. Ap.J., 208, 346-53.

Herbst, W. & Assousa, G.E. (1978). The Role of Supernovae in Star Formation and Spiral Structure. In Protostars & Planets, ed. T. Gehrels, 368-83, University of Arizona Press, Tucson.

Kerr, F.J. (1977). Star Formation and the Galaxy. In Star Formation, ed. T. de Jong & A. Maeder, 3-36, D. Reidel, Dordrecht.

Kormendy, J. & Norman, C.A. (1979). Observational Constraints on Driving Mechanisms for Spiral Density Waves. Ap.J., 233, 539-52.

Lada, C.J., Blitz, L. & Elmegreen, B.G. (1978). Star Formation in OB Associations. In Protostars & Planets, ed. T. Gehrels, 341-67, University of Arizona Press, Tucson.

Larson, R.B. (1977). Rates of Star Formation. In The Evolution of Galaxies and Stellar Populations, ed. B.M. Tinsley & Larson, R.B., 97-121, Yale University Printing Service, New Haven.

Larson, R.B. & Tinsley, B.M. (1978). Star Formation Rates in Normal and Peculiar Galaxies. Ap.J., 219, 46-59.

Lequeux, J. (1980). Empirical Information on Star Formation in Galaxies. In Star Formation, ed. A. Maeder & L. Martinet, 77-132, Geneva Observatory, Sauverny.

Lin, C.C. & Shu, F.H. (1964). On the Spiral Structure of Disk Galaxies. Ap.J., 140, 646-55.

Lynds, R. & Toomre, A. (1976). On the Interpretation of Ring Galaxies : The Binary Ring System II. Hz 4. Ap.J., 209, 382-88.

Madore, B.F. (1977). Numerical Simulations of the Rate of Star Formation in External Galaxies. M.N.R.A.S., 178, 1-9.

Madore, B.F. (1978). Supergiants, Spiral Structure and Star Formation in M33. Observatory, 98, 169-72.

Madore, B.F. (1980). Companions to Nearby Spirals. A.J., 85, 507-12.

Madore, B.F. (1981). Galaxy-Induced Star Formation. (in preparation).

Mezger, P.G. & Smith, L.F. (1977). Radio Observations Related to Star Formation. In Star Formation, ed. T. de Jong & A. Maeder 133-78, D. Reidel, Dordrecht.

Miller, G.E. & Scalo, J.M. (1979). The Initial Mass Function and Stellar Birthrate in the Solar Neighborhood. Ap.J. Suppl., 41, 513-47.

Miller, R.H. & Smith, B.F. (1979). Formation of an Elliptical Galaxy. In Photometry Kinematics and Dynamics of Galaxies, ed. D.S. Evans, 365-8, Department of Astronomy, Austin.

Mueller, M.W. & Arnett, W.D. (1976). Propagating Star Formation and Irregular Structure in Spiral Galaxies. Ap.J., 210, 670-8.

Newton, K. (1980). Neutral Hydrogen and Spiral Structure in M33. M.N.R.A.S., 190, 689-710.

Noerdlinger, P.D. (1979). Binary Galaxy Statistics. III. Correlations in Angular Measures, Sense of Rotation, and Type. Ap.J., 229, 877-90.

Opik, E.J. (1953). Stellar Associations and Supernovae. Irish Astron.J., 2, 219-33.

Oort, J. (1954). Outline of a Theory on the Origin and Acceleration of Interstellar Clouds and OB Associations. B.A.N., 12, 177-86.

Page, T. (1975). Binary Galaxies. In Stars and Stellar Systems, 9, ed. A.R. Sandage, M. Sandage & J. Kristian, 541-56, University of Chicago Press, Chicago.

Roberts, W.W. (1969). Large-Scale Shock Formation in Spiral Galaxies and its Implications on Star Formation. Ap.J., 158, 123-43.

Rots, A.H. & Shane, W.W. (1975). Distribution and Kinematics of Neutral Hydrogen in the Spiral Galaxy M81 I. Observations. Astron. Astrophys., 45, 25-42.

Sandage, A.R. (1963). Photoelectric Observations of the Interacting Galaxies VV117 and VV123 Related to the Time of Formation of Their Satellites. Ap.J., 138, 863-72.

Scalo, J.M. (1978). The Stellar Mass Spectrum. In Protostars & Planets, ed. T. Gehrels, 265-87, University of Arizona Press, Tucson.

Schmidt, M. (1959). The Rate of Star Formation. Ap.J., 129, 243-58.

Schmidt, M. (1962). The Evolution of the Sun's Neighborhood. In Symposium on Stellar Evolution, ed. J. Sahade, 67-73, Observatorio Astronomico, La Plata.

Schmidt, M. (1963). The Rate of Star Formation. II. The Rate of Formation of Stars of Different Mass. Ap.J., 137, 758-69.

Seiden, P.E., Schulman, L.S. & Gerola, H. (1979). Stochastic Star Formation and the Evolution of Galaxies. Ap.J., 232, 702-5.

Seiden, P.E. & Gerola, H. (1979). Properties of Spiral Galaxies from a Stochastic Star Formation Model. Ap.J., 233, 56-66.

Sharp, N. & Jones, B. (1980). Peculiarities of Binary Galaxies. Nature, 283, 275-7.

Talbot, R.J. & Arnett, W.D. (1975). The Evolution of Galaxies. IV. Highly Flattened Disks. Ap.J., 197, 551-70.

Talbot, R.J. (1977). Evolution of Galaxies Governed by Disk Dynamics and Spiral Structure. In Chemical and Dynamical Evolution of our Galaxy, ed. E. Basinska-Grzesik & M. Mayor, 31-46, Geneva Observatory, Sauverny.

Tomov, A.N. (1978). UBV Observations of Double Spiral Galaxies, Sov. Astron., 22, 540-4.

Toomre, A. (1969). Group Velocity of Spiral Waves In Galactic Disks. Ap.J., 158, 899-913.

Toomre, A. (1978). Interacting Systems. In IAU Symp. No. 79, The Large-scale Structure of the Universe. ed. M.S. Longair & J. Einasto, 109-16, Reidel, Dordrecht.

White, S.D.M. & Valdes, F. (1980). The Luminosity Function of Close Binary Galaxies. M.N.R.A.S., 190, 55-70.

GLOBULAR CLUSTER SYSTEMS: IMPLICATIONS FOR GALAXY FORMATION

K.C. Freeman

Mount Stromlo and Siding Spring Observatories
Research School of Physical Sciences
Australian National University

This is a review of some properties of globular cluster systems in galaxies (including our own) that bear on our galaxy formation picture. Globular cluster systems are the most distended luminous component of galaxies, and so are probably least affected by dissipational processes during galaxy formation. Also they include objects with a range of chemical abundances, from very low (but not <u>the</u> lowest in the Galaxy) to intermediate. By looking at cluster ages, orbital properties and abundances, we have the opportunity to build up an empirical picture of the early collapse and chemical evolution of galaxies.

GLOBULAR CLUSTER POPULATIONS

Harris and Racine (1979) estimated the total number of globular clusters in a sample of elliptical and spiral galaxies. This total number was derived via the luminosity function for globular clusters, which is apparently fairly uniform from galaxy to galaxy. For most ellipticals, the total number is closely proportional to the integrated luminosity of the galaxy. An exception is M87: it has about the same integrated luminosity as NGC 4472, but about four times as many clusters. A similar difference in cluster population is seen between the two brightest ellipticals in the Fornax cluster.

The spiral galaxies appear to have fewer clusters per unit luminosity than do the ellipticals. However Harris and Racine, and Kormendy (personal communication) suggest that the number of clusters per unit <u>bulge</u> luminosity is close to that for the ellipticals.

NGC 5128 remains discrepant. It has a very large elliptical or bulge component, and very few globular clusters (van den Bergh, 1980).

K.C. FREEMAN

GALACTIC DISTRIBUTION OF GLOBULAR CLUSTERS

The review by Harris and Racine (1979) covers this subject fully. In our galaxy, the globular cluster system is roughly spherical, with maybe some flattening of the innermost metal rich subsystem. The radial distribution of the surface density of the clusters in the Galaxy follows the $R^{\frac{1}{4}}$ law closely, and this is seen again in M31. The analysis by de Vaucouleurs and Buta (1978) of the catalogue by Sargent et al (1977) for the M31 clusters showed how the clusters and the diffuse spheroidal bulge both follow the $R^{\frac{1}{4}}$ law, with a common scale length. So it is tempting to identify the cluster system with this spheroidal bulge. For elliptical galaxies, the cluster distributions are also well represented by the same law, and again it seemed reasonable to identify the cluster system dynamically with the luminous diffuse component. However it seems now that this convenient picture may not be correct. In a recent review, Racine (1980) argues, from a critical review of the data, that the globular cluster systems are significantly more distended than the underlying light. In M87, for example, the luminous volume density follows a radial R^{-3} law, while the cluster number density is closer to $R^{-2.4}$. So we are now tempted to identify globular cluster systems as dynamically intermediate between the dark matter and the luminous bulge or elliptical.

ABUNDANCE GRADIENTS IN GLOBULAR CLUSTER SYSTEMS

This remains a rather confused subject. For the Galactic system, it has long been known that the innermost clusters have a higher mean chemical abundance than the outer clusters. The important question is whether there is a monotonic gradient throughout the system, with only very metal weak clusters in the outermost parts. Opinions differ. Searle and Zinn (1978) and Zinn (1980) find no evidence for an abundance gradient with radius R or height z above the galactic plane, for clusters with R between 9 and 40 kpc. Clusters in this zone have a wide range of abundance, from -1.5 to -2.3 in [Fe/H] . Harris and Canterna (1979), on the other hand, include clusters out to R = 100 kpc in their discussion. Although there are only a few of these outer clusters, they make an important contribution to the apparent trend of abundance with radius, and Harris and Canterna argue that both the range of abundance and the mean abundance decrease continously with increasing radius.

Recently there has been a suggested revision of the abundance scale (see Cohen, 1980, and Pilachowski, 1980). The effect of this revision is to reduce the abundance values for metal rich clusters like M71 and 47 Tuc, from about -0.5 to about -1.2. This revision is not yet accepted by all workers in this field. Its major effect on the largescale abundance structure of the globular cluster

system is to reduce the mean abundance and range of abundance in
the inner parts of the galaxy (R < 8 kpc) where most of the metal
rich clusters are found. With this new abundance scale, the
abundance gradient in the inner parts almost disappears. The
radial abundance distribution is then almost uniform for all
R < 40 kpc (see Zinn, 1980). I will use the old scale for the
rest of this review.

Another contribution to this unclear picture of the abundance
gradient in the galactic halo comes from a study of the abundances
of RR Lyrae stars near the North Galactic Pole, by Butler et al
(1979). These stars show no evidence for an abundance gradient
with z, out to z = 25 kpc. As for the globular clusters, there
is a fairly wide range of abundance at all z, from about −1.0 to
−2.2 in [Fe/H] . Butler et al point out that the absence of RR
Lyraes with [Fe/H] < −2.2 is probably a stellar evolution effect.
Lower abundance populations do not populate the part of the horizontal branch where the RR Lyraes are found. This means that the
RR Lyraes could give us a rather biased picture of the abundance
distribution in the galactic halo. This is reinforced by the
discovery of a few very metal weak giants, with [Fe/H] < −2.7.
We should take the same point for the globular cluster system.
No clusters are known with [Fe/H] < −2.2, but this does not mean
that a lower abundance stellar population does not exist. When
the abundance distribution for halo giants is known, it may well
turn out that the galactic abundance gradient is more apparent
than it seems to be for the globular clusters and RR Lyrae stars.

For the globular cluster system of M87, the abundance gradient
is more clearly established. Strom et al (1980) have made UBR
photometry of 1700 clusters in this galaxy, over the radius
interval 9 to 54 kpc. The radial gradient of (U−R) colors for
the clusters is fairly clear. It has a similar lengthscale to
the gradient in (U−R) for the diffuse light. But, at a particular
radius, the clusters are about 0.5 mag bluer, in the mean, than
the diffuse light. This is presumably the result of a local
abundance difference between the clusters and the diffuse stellar
component. The size of this local abundance difference is about
0.6 dex in [Fe/H] . Here is another fairly firm indication that
the cluster system and the diffuse bulge or elliptical should not
be too closely identified. (The first indication was the difference
in lengthscale for their radial distributions). Strom et al make
another interesting point, in comparing the Galactic and M87
cluster systems. When mean cluster colors are displayed against
R/R_e (R_e is the effective or half light radius), the color gradients
are similar for the two systems, and the local spread in color at
each value of R/R_e is also similar. (For the Galaxy, R_e = 3.3 kpc,
and this comparison is made out to R/R_e = 5). Strom et al argue
that this observation is against the formation of ellipticals by
mergers of fully formed spirals. To form M87 by such mergers,
ten spirals like ours would be required, and these mergers should

reduce the relative color gradient and increase the local scatter.

Returning now to the Galactic globular cluster system, we recall that Zinn's sample of clusters (which are on a uniform abundance system) showed no abundance gradient for clusters now in the zone with R between 9 and 40 kpc, when the abundance is displayed against the galactocentric distance R. However, if there is an underlying abundance gradient in the cluster system, the galactocentric distance may not be the most appropriate coordinate against which to see it. For example, say the clusters formed near their orbital perigalactic radius R_{min}. Then an abundance gradient would probably be most evident in the abundance - R_{min} plane. In the abundance - R plane, the gradient would be smoothed out by the distribution of R/R_{min} over time for each orbit.

It seemed worth checking this point. In Zinn's sample of clusters with R between 9 and 40 kpc, there are 25 clusters with measured tidal radii. From these tidal radii, it is possible to estimate the perigalactic radius R_{min} (see the discussion later in this review). It turns out that these 25 clusters show a very clear abundance gradient in the abundance - R_{min} plane. Clusters with large values of R_{min} have low metal abundances. It appears that the more metal rich clusters found now in the zone with R between 9 and 40 kpc are there because they are on highly eccentric orbits with small perigalactic distances. See also Harris and Canterna (1979). This may be a hint that the clusters did form near perigalacticon. It also foreshadows an unexpected effect that will be discussed later.

CLUSTER AGES AND THE SECOND PARAMETER

Demarque (1980) has given a very useful review of this topic. Several authors have derived ages for globular clusters, by fitting theoretical isochrones in the M_V - (B-V) plane to the observed color-magnitude diagrams that extend below the main sequence turnoff. Different authors use different fitting procedures, which produces a spread of derived ages for a particular cluster. Nevertheless, on the system of the Yale isochrones, (Ciardullo and Demarque 1977), there appears to be a clear trend of abundance with age for the nearby globular clusters, going from [Fe/H] = -2.2 at 16 Gyr to [Fe/H] = -0.5 at 10 Gyr. This represents a fairly slow chemical enrichment of the globular cluster system. Because this age scale is tied to the standard stellar models, there is some additional uncertainty associated with the reliability of the models themselves. Internal core rotation and internal mixing, for example, are not properly understood; they would probably affect the derived age scale if they are significantly present. For comparison, the oldest clusters of the old disk have ages of about 6 Gyr.

The morphology of the horizontal branch is sensitive to age and to chemical abundance. The population of the horizontal branch becomes bluer with increasing age at a given abundance, or with decreasing abundance at a given age. For the metal weak clusters ($[Fe/H] < -1.5$) in the inner parts of the Galaxy ($R < 5$ kpc), the horizontal branch stars are almost all on the blue side of the RR Lyrae gap. With increasing distance from the galactic center, the horizontal branches of these metal weak clusters become progressively more diverse: some include significant numbers of stars on the red side of the RR Lyrae gap. Another parameter (the second parameter), in addition to the metal abundance, is affecting the morphology of the horizontal branch in these outer clusters. Age, $[CNO/Fe]$ and $[He/H]$ are three prominent possibilities. Searle and Zinn (1978) and Zinn (1980) argue that age is the second parameter. If this is correct, then the data suggest that there is a downward age spread of about 4 Gyr among the metal weak clusters beyond R = 12 kpc, relative to those with $R < 5$ kpc. This in turn suggests that chemical enrichment proceeded more slowly in the outer parts of the Galaxy. See Demarque (1980) for a fuller discussion.

THE ABUNDANCE DISTRIBUTION FOR GALACTIC GLOBULAR CLUSTERS

Searle and Zinn (1978) gave an interesting discussion of the abundance distribution f(z) for the Galactic globular cluster system. (z is now the linear abundance and f(z) is its probability density). Clusters in the region beyond R = 9 kpc, where the abundance gradient is not obvious, show a roughly exponential distribution over abundance. Searle and Zinn compare this distribution with the one predicted by the simple model of chemical evolution. For a homogeneous closed system evolving to completion (i.e. all the gas is used up), the predicted f(z) has the form

$$f(z) = (1/\langle z \rangle) \exp(-z/\langle z \rangle)$$

where $\langle z \rangle$ is the average abundance for stars formed. This is an excellent fit to the observed distribution for the outer clusters. If one were to take this model seriously, it would mean that the gas was completely converted into stars during the halo formation phase, with the yield equalling the mean z for the halo stars. Then the galactic disk would have formed from mass lost from the evolving halo stars. This seems an unlikely story. However it turns out that the exponential form of f(z) given above is also predicted by a model for which the gas is used up only partly by star formation, the remainder being lost from the star formation region, as suggested by Hartwick (1976). So Searle and Zinn proposed that the chemical evolution of the halo proceeds in fragments, which lose gas at a rate proportional to the rate at which stars form (because of SN disruption, for example).

Such fragments evolving to completion will again produce the exponential f(z), even though only a fraction of the fragment mass is converted into stars. The mean abundance $\langle z \rangle$ is then the effective yield

$$y_{eff} = y/(1+a)$$

where y is the yield in the absence of mass loss, and a is the ratio of gas loss rate to gas consumption rate by star formation. So a larger value of a produces a lower mean halo abundance, and relatively less mass locked up in halo stars. This makes sense: M31 has a higher mean abundance for its halo globular clusters and a larger bulge to disk ratio than the Galaxy (Searle 1978).

GLOBULAR CLUSTER INITIAL MASS FUNCTIONS

Da Costa (unpublished) studied the main sequence mass functions of three nearby globular clusters. Over the mass interval 0.4 to 0.8 solar masses, he derived the slope x of the mass function (x = 1.35 for the Salpeter function). Interpreted as the slope of the initial mass function, he found that x changed from 0.9 ± 0.5 for NGC 6397 ([Fe/H] = -2.2) to 2.9 ± 0.5 for 47 Tuc ([Fe/H] = -0.6), where the uncertainties include the effects of evaporation and tidal shocks. This is a hint that the IMF may have been flatter (i.e. more massive stars) when the metal abundance was lower. However Illingworth's (unpublished) results for M92 ([Fe/H] = -2.2) suggest that its mass function is fairly steep.

ORBITAL PROPERTIES OF GLOBULAR CLUSTERS

If we want to know how the formation of the globular cluster system fits in to our overall picture of galaxy formation, then it seems essential to know how the kinematical (or orbital) properties of the clusters relate to their chemical properties. This kind of data, for the subdwarfs in the solar neighborhood, for example (Eggen et al 1962), has been very important in the evolution of our present picture of galaxy formation.

One approach to learning about the orbital properties of the galactic globular clusters is via kinematical solutions like those of Frenk and White (preprint). They use the position and radial velocity data for each cluster to show that the cluster system has a systematic net rotation of about 60 km/s and that the velocity dispersion is fairly close to isotropic.

Another approach is to use the tidal radius of the cluster. Assume this is defined by the galactic tidal field at perigalact-

icon. If the tidal field is known as a function of radius, and
the mass of the cluster is derived from its integrated luminosity,
then the perigalactic distance can be estimated. The ratio of
perigalactic distance to present distance from the galactic center
is a statistical estimate of the orbital eccentricity (Peterson
1974; Racine and Harris 1975). There are certainly problems
with this approach. For example, it is not entirely clear yet
how the tidal radius of the cluster and the tidal field of the
galaxy are related, and the effects of tidal shocks may be significant in defining the tidal radius for some clusters. Also the
sample of clusters with accurate tidal radii is still relatively
small. With these cautions, I will now describe briefly some
recent work with Pat Seitzer. This led to an unexpected result
that is closely associated with the correlation of cluster
abundance with perigalactic radius, that was mentioned earlier.

First we assume that the galactic potential is spherically
symmetric and has the logarithmic form associated with a flat
rotation curve (see Blitz 1979 for a recent composite galactic
rotation curve) with $V = 250$ km/s. The cluster masses are
derived from their luminosities, using $M/L = 2$. We can then
estimate the ratio R_{min}/R for each cluster, from its tidal radius;
R_{min} is again the perigalactic distance for the cluster orbit,
and R is its present galactocentric distance. If we define
the orbital eccentricity in the usual way as

$$e = (R_{max} - R_{min})/(R_{max} + R_{min})$$

where R_{max} is the apogalactic distance, then the expectation
value of the ratio R_{min}/R is $(1 - e)/(1 + 0.42e^2)$, so R_{min}/R
is an estimate of the orbital eccentricity. For a sample of
about 50 clusters with fairly well determined tidal radii, two
results were apparent.

(i) there was no obvious trend of R_{min}/R with R itself,
which argues against the view that the outer clusters are in
highly elongated orbits.

(ii) there is a fairly clear trend of R_{min}/R with abundance,
for [Fe/H] < -1.0. The trend is in the sense that the most
metal weak clusters are in the least eccentric orbits. (See
Racine and Harris 1975 for a somewhat similar conclusion). This
is certainly not what we would have expected to find for the
cluster system, if we believe that the halo subdwarfs in the
solar neighborhood and the globular clusters belong to the same
population. It is probably another hint that we should not be
too ready to associate the cluster system with the metal weak
diffuse halo. At least we can see now why the unmistakeable
abundance gradient in the abundance – R_{min} plane becomes very
difficult to see in the abundance – R plane.

This trend, of increasing orbital eccentricity with increasing abundance, could be quite important for our picture of how the globular cluster system formed. Because of the possible problems associated with using the cluster tidal radii that I mentioned earlier, it would be reassuring to have an independent check. This comes from the M31 globular clusters, for which van den Bergh (1969) has measured radial velocities and estimated abundances. We can at least check whether their kinematics are consistent with the possibility that the low abundance clusters are in low eccentricity orbits. This is easily done. M31 has a flat rotation curve, and we can calculate the expected frequency distributions of observed radial velocity for collections of randomly orientated orbits of low and of high eccentricity in a flat rotation curve potential. Fortunately these theoretical frequency distributions are quite different. The low eccentricity orbits have an almost uniform distribution of observed radial velocity, over the interval −V to +V, where V is the circular velocity. The high eccentricity orbits have a distribution that is strongly peaked around the systemic velocity.

It turns out that the metal weak clusters in M31 ($L \leq 6$ on van den Bergh's scale) do show a nearly uniform distribution of radial velocity, and the intermediate abundance clusters ($6 < L < 10$) have a peaked distribution. Although the number of clusters is not large, the two observed distributions are different at the 95 percent confidence level, and they suggest again that the metal weaker clusters in M31 are in lower eccentricity orbits. Full details of this work will be published soon.

CONCLUSION

I think that the important conclusion that comes out of all this, from the radial distributions, colors and orbital properties of the globular cluster systems, is that the cluster systems and the diffuse outer halo populations of galaxies are in many ways different, and so probably had a different chemical and dynamical history. The cluster systems may well have formed rather earlier.

REFERENCES

Blitz, L.(1979). The rotation curve of the Galaxy to R = 16 kiloparsecs. Ap.J. 231, L115–119.

Butler, D., Kinman, T.D. & Kraft, R.P. (1979). Metal abundances of RR Lyrae variables in selected galactic star fields. II. The Lick astrograph fields near the North Galactic Pole. Astron. J. 84, 993–1004.

Ciardullo, R.B. & Demarque, P. (1977). Yale Trans. 33.

Cohen, J.G. (1980). Chemical properties of individual globular clusters. In IAU Symposium 85, ed J. Hesser. Reidel. pp385-394.

Demarque, P. (1980). Ages and abundances of globular clusters and the oldest open clusters. In IAU Symposium 85, ed J. Hesser. Reidel. pp281-296.

de Vaucouleurs, G. & Buta, R. (1978). On the distribution of globular clusters in the Galaxy and M31. Astron. J. 83, 1383-1389.

Eggen, O.J., Lynden-Bell, D. & Sandage, A. (1962). Evidence from the motions of old stars that the Galaxy collapsed. Ap.J. 136, 748-766.

Harris, W.E., & Canterna, R. (1979). On the abundance gradient in the galactic halo. Ap.J. 231, L19-23.

Harris, W.E. & Racine, R. (1979). Globular clusters in galaxies. In Annual Reviews of Astronomy and Astrophysics, 17, 241-274.

Hartwick, F.D.A. (1976). The chemical evolution of the galactic halo. Ap.J. 209, 418-423.

Peterson, C. (1974). Distribution of orbital eccentricities of the globular clusters. Ap.J. 190, L17-20.

Pilachowski, C.A., Sneden, C. & Canterna, R. (1980). Chemical Composition of Southern globular clusters. In IAU Symposium 85, ed J. Hesser. Reidel. pp467-468.

Racine, R. (1980). Globular cluster systems in galaxies. In IAU Symposium 85, ed J. Hesser. Reidel. pp369-380.

Racine, R. & Harris, W.E. (1975). A photometric study of NGC 2419. Ap.J. 196, 413-432.

Sargent, W.L.W., Kowal, C.T., Hartwick, F.D.A. & van den Bergh, S. (1977). Search for globular clusters in M31. I. The disk and the minor axis. Astron. J. 82, 947-953.

Searle, L. (1978). Presentation at NATO Advanced Study Institute on Globular Clusters, Cambridge, England. (Unpublished).

Searle, L. & Zinn, R. (1978). Composition of halo clusters and the formation of the galactic halo. Ap.J. 225, 357-379.

Strom, S.E., Forte, J.C., Harris, W.E., Strom, K.M., Wells, D.C. & Smith, M.G. (1980). Preprint.

van den Bergh, S. (1969). Photometric and spectroscopic observations of globular clusters in the Andromeda nebula. Ap.J. Suppl. 19, 145-174.

van den Bergh, S. (1980). Star clusters as touchstones of theories of galactic evolution – a few examples. In IAU Symposium 85, ed J. Hesser. Reidel. pp1-8.

Zinn, R. (1980). Preprint.

GLOSSARY OF ABBREVIATIONS AND

TECHNICAL TERMS

A2 p. 192   The second Ariel satellite.

Boltzmann Equation p. 55 see Collisionless Boltzmann Equation.

C1 Chondrites p. 214   A type of Carbonacious Chondrite meteorite
of low specific gravity ~2.2 and with a high content
of water ~20%, sulphur ~6% and carbon ~3.5%.

Classical Integral p. 44   see Integral.

Collisionless Boltzmann Equation pp. 55, 63 Conservation of mass
tells us that the decrease of mass in any volume τ of
the 6 dimensional phase space of positions and velocities is due to stars moving through the walls. The
stars change their positions $\underline{r}$ and velocities $\underline{v}$ according to the equations

$\underline{\dot{r}} = \underline{v}$, $\underline{\dot{v}} = -\partial\Phi/\partial\underline{r}$ where Φ is the gravitational

potential so conservation of mass tells us that for
any volume

$$-\frac{\partial}{\partial t}\int_\tau f d\tau = \int_S f(\underline{v}, -\frac{\partial\Phi}{\partial\underline{r}}) \cdot d\underline{S}$$

where f is the distribution function and S is the
surface of τ. Applying the divergence theorem and
using the fact that τ is arbitrary, we find that for
any volume $d\tau$

$$\left[\frac{\partial f}{\partial t} + \frac{\partial}{\partial\underline{r}} \cdot (f\underline{v}) - \frac{\partial}{\partial\underline{v}} \cdot (f\frac{\partial\Phi}{\partial\underline{r}})\right]d\tau = 0$$

Now $\underline{r}$ and $\underline{v}$ are independent coordinates in phase space
so $\frac{\partial u}{\partial x}$ etc are all identically zero and since Φ is
independent of $\underline{v}$ we find

$$\frac{Df}{Dt} \equiv \frac{\partial f}{\partial t} + \underline{v} \cdot \frac{\partial f}{\partial\underline{r}} - \frac{\partial\Phi}{\partial\underline{r}} \cdot \frac{\partial f}{\partial\underline{v}} = 0$$

which is the Collisionless Boltzmann equation. Above
we have introduced the notation D/Dt for the convec-
tive derivative that follows the motion in phase space.
This is analogous to the operator $\frac{\partial}{\partial t} + \underline{u} \cdot \underline{\nabla}$ in fluid

261

mechanics but is defined by

$$\frac{D\,Q}{Dt} = \lim_{\delta t \to o} \frac{Q\,(\underline{r} + \delta\underline{r},\ \underline{v} + \delta\underline{v},\ t + \delta t) - Q(\underline{r}, \underline{v}, t)}{\delta t}$$

where $\delta\underline{r} = \underline{v}\ \delta t$ and $\delta\underline{v} = -\frac{\partial \Phi}{\partial \underline{r}}\ \delta t$.

Thus the solution to the Collisionless Boltzmann equation is any function f that does not change when one "moves with the stars" around phase space.

Distribution function f pp. 44, 59, 60 f(x,y,z,u,v,w,t) dx dy dz du dv dw is the total mass of those stars that lie in the box defined by coordinates in the ranges x, x+dx; y, y+dy; z, z+dz; which simultaneously have velocity components in the ranges u to u+du, v to v+dv, w to w+dw. The density ρ is related to the distribution function f by

$$\rho(\underline{r}) = \int f(\underline{r},\ \underline{v},\ t)\ du\ dv\ dw$$

H.E.A.O. pp. 191, 192 High Energy Astronomical Observatory satellites A & B.
H.E.A.O.B. = Einstein Observatory.

I.G.M. pp. 211, 213 Inter-Galactic Medium. The gas between the galaxies in a cluster of galaxies.

I.M.F. pp. 219, 221 Initial Mass Function. This function φ(m)dm gives the number of stars born with masses between m and m+dm. It is often for simplicity assumed to be the same everywhere even though there is little direct evidence for this.

Integral pp. 44, 48, 60 short for "first integral of the equations of motion" or "constant of the motion". These arise in stellar dynamics as follows. The collisionless Boltzmann equation can be written Df/Dt = o so f does not change following the motion along the trajectories of the stars through phase-space. For a steady state system this implies that f is constant along those trajectories at all times. Now functions of position and velocity that do not change along the trajectory of any star are called "constants of the motion" (or sometimes as "first integrals of the equations of motion"). Hence in any steady system f must be a constant of the motion and can be expressed as a function of the constants of motion used to define the trajectories. Thus we have Jeans's theorem that

the distribution function f must be a function of
"integrals".

A star needs six numbers to define its initial
position and velocity. A one dimensional trajectory
will require one less; so five independent integrals
of the motion are sufficient to define a trajectory
in phase-space.

Integral, classical p. 44   In an axially symmetrical galaxy,
both the specific energy, ε, the energy per unit mass,
and the specific angular momentum about the axis are
first integrals. However, as the star moves in the
galaxy it oscillates both radially and perpendicularly
to the galactic plane. In practice, the amplitudes of
these two oscillations do not vary over all the values
allowed by energy and angular momentum conservation.
Rather there are new variables whose oscillations are
excited independently so that their amplitudes are
independently constant. If we take the amplitude of
one of these to be a third integral I_3 then the amplitude of the other is determined by energy conservation.

One may well ask why stop at 3 integrals and
there is no complete answer to this question. Systems
are known, such as motion in the field of a point mass,
in which there are five classical integrals and all
are in general necessary. However, there is a rough
answer as follows. The three dimensional motion of a
star is a superposition of three basic movements, an
oscillation in and out, an oscillation up and down,
and galactic rotation. Except in very special galaxy
potentials (such as spherical ones) the frequencies of
these oscillations are not directly related, and so in
general their ratios are irrational. Thus we expect
the three phases of those oscillations to have no
regular relationship. Under these conditions a single
star's trajectory in phase space fills, not a line,
but a three dimensional volume, since any phase of one
oscillator occurs arbitrarily close to any phase of
another. Thus, in most stellar systems we expect three
effective integrals corresponding to the three independent amplitudes of the oscillations, and their values
will define the 3 volumes of phase space.

This expectation, although generally fulfilled,
breaks down for special cases of high symmetry when
there are more, effective integrals and sometimes for
less regular potentials or at large oscillation amplitudes close to two or more resonances. The phase space
trajectories in the latter cases may fill a four

263

dimensional sub-region of phase space, or even a five dimensional one. Orbits that appear to have this behaviour are called stochastic.

I.R.   pp. 223, 227   Instantaneous Recycling. Since the stars that make metals evolve on a $10^6 - 10^7$ year timescale, this is very short compared to the age of the Galaxy. In the instantaneous recycling approximation, this star evolution time is neglected so that each star formation episode is followed immediately by supernovae that enrich the interstellar gas. The approximation is very useful as it replaces integro-differential equations by simple differential equations.

I.S.M.   pp. 24, 213, 214   Inter-Stellar Medium. The gas and dust between the stars inside a galaxy.

S Process   p. 226   Slow neutron capture process. Element transformation by a flux of neutrons applied sufficiently slowly that there is time for β unstable nuclei to decay between one neutron capture and the next.

V.L.A.   p. 174   Very Large Array of twenty seven, eighty two feet radio telescopes built near Socorro, New Mexico.

Yield   p. 225   The yield of a particular element, iron say, is the mass of iron returned to the interstellar medium after being newly created in a generation of stars, divided by the mass of all the stars formed in that generation.

INDEX OF ASTRONOMICAL OBJECTS

269

INDEX